INDIA: ECONOMIC, POLITICAL AND SOCIAL ISSUES

INDIA

ENVIRONMENTAL, POLITICAL AND SOCIAL ISSUES

INDIA: ECONOMIC, POLITICAL AND SOCIAL ISSUES

Additional books and e-books in this series can be found on Nova's website under the Series tab.

INDIA: ECONOMIC, POLITICAL AND SOCIAL ISSUES

INDIA

ENVIRONMENTAL, POLITICAL AND SOCIAL ISSUES

JULES LESSARD
AND
ORSON RIVIÈRE
EDITORS

nova
science publishers
New York

Library of Congress Cataloging-in-Publication Data

ISBN: 978-1-53618-413-6

Published by Nova Science Publishers, Inc. † New York

CONTENTS

PREFACE

India: Environmental, Political and Social Issues first highlights the environmental conditions of Uttarakhand such as relief, structure, drainage, climatic conditions and forest types, as well as the influence of these aspects on aspects of demographic structure such as population distribution, growth, sex ratio and literacy.

The authors go to present findings indicate that the Tropical Lowland Rain Forests in Sri Lanka are capable of recovery, in terms of biodiversity and stature, if sufficient recess is provided following land use practices such as selective logging and chena cultivation in spatial and temporal contexts.

As a desert state, the climate of Rajasthan is uniquely characterized by low rainfall, dryness and extreme temperature. Fauna of this region are specifically adapted to this hostile environment. In this compilation, threats to fauna of Rajasthan are assessed.

Later, after presenting a short introduction to the historical and cultural background, the authors narrate Karunanidhi's most decisive film, Parasakthi (Supreme Power, 1952). It challenges certain cultural stereotypes such as the sexual exploitation of women, social discrimination, corruption, religious superstition and so on.

The closing study examines the effects of socio-economic determinants on child malnutrition in Uttar Pradesh, the most populous and malnourished state of India. Using data from NFHS-4, this study highlights district-wide variations in the nutritional status of children less than five years old in the state.

Chapter 1 - Historically, Uttarakhand is considered as densely forested area, place of God and Goddesses and the home of wild animals and also got recognition as a confluence of various cultural groups came from Central Asia in the north, Nepal in the east and Himachal in the west and the Gangetic Plain in the south. Geographically it extends from the Tons - feeder of Yamuna River in the west to the Kali or Sharda in the east. Its northern boundary is demarcated by Indo - Tibet water parting ridge separating it from Tibet (China) and southern boundary corresponds almost with the southern limit of Tarai belt separating it from Uttar Pradesh. Situated in the Himalayan Region, all the districts except Hardwar and U. S. Nagar of the Plain, Tarai and Bhabar region are mountainous tracts. Historically, the region under the present study has been divided into two divisions- Garhwal (Kedarkhand) and Kumaun (Manaskhand) – presently Garhwal includes the districts of Hardwar, Dehradun, Uttarkashi, Tehri, Rudraprayag, Chamoli and Pauri, and Kumaun spread over the districts of Almora, Bageshwar, Pithoragarh, Champawat, Nainital and U. S. Nagar. About 85% of the land area of the region is mountainous and remaining fall in the almost plain area- Bhabar, duns, Tarai and river Ganga plain. Lying between $28^{0}44'$to $31^{0}25'$ north latitudes and $77^{0}45'$ to $81^{0}1'$ east longitudes, it encompasses an area of 53483 km^2 and supports 10086292 people in 2011. Due to the unfavorable physical conditions only 49% population is made their habitat in the mountainous region of the Uttarakhand state. Out of the total population 69.8% (7036954 persons) are residing in rural villages and the remaining 30.2% (3049338 persons) people are registered as urban population. Out of the total population 18.8% and 2.89% population belongs to scheduled caste and scheduled tribe population respectively while in rural Uttarakhand these categories have respectively 21.3% and 3.76% population. The average sex ratio in Uttarakhand is 963 females per thousand males. It is noticed that the sex ratio among children (0-6 years), scheduled caste and the scheduled tribe is less than the total population in these respective categories. The average literacy rate in Uttarakhand is 78.8%. The average growth rate of the region is 18.8% which is more than the country average. The main aim of this study is to highlight the physical environmental conditions such as relief, structure, drainage, climatic conditions, forest types and influence of these aspects on demographic structure such as population distribution, growth, sex ratio, literacy, etc of Uttarakhand. The present study is based on secondary data released by the Census of India, 2011 and literature available in the public domain. The findings of this study will be helpful to make a feasible plan for the all-round sustainable development of Uttarakhand.

Chapter 2 - As nature, the forest too is in a state of flux and it possesses innate ability to thwart impacts, both natural and anthropogenic. The majority of the Tropical Rain Forests in Sri Lanka have suffered selective logging, more intensely five to six decades ago. The Sinharaja and Kanneliya are two such largest remaining lowland rain forest areas in Sri Lanka. The National Conservation Review (NCR) that was conducted between 1991 and 1996 to evaluate all remaining natural forests used Gradient-directed transect (gradsect) method to sample these forest blocks to assess species diversity. The Biodiversity Monitoring Survey (BMS) was conducted in 2008 to determine the changes in the status of plant biodiversity in Sinharaja and Kanneliya forests since the NCR survey was conducted in these sites in 1992, i.e., over sixteen years prior to the BMS. It was evident that Kanneliya, along with two other contiguous forest areas, possess richest plant diversity with highest number of species and endemic species in the island. Positive trends in some important biodiversity characteristics after sixteen years were noted. Changes in succession status in both sites, i.e., increase in primary species and decrease in combined pioneer and secondary species clearly indicated that forests are successfully regenerating accompanied by an improvement in biodiversity quality. In Sinharaja the recovery in sites that were abandoned after slash and burn (chena) cultivation was clearly evident by similar observations. In Kanneliya, after nearly two decades of recess since timber exploitation, observations indicated recovery of forest stature accompanied by the appearance of emergents and reproductively mature canopy trees gradually bringing back the original seed bank into the forest dynamics. The enrichment planting with Mahogany in selectively logged plots in Kanneliya during early 1990's was a proven failure. The above findings indicated that the Tropical Lowland Rain Forests are capable of recovery, in terms of biodiversity and stature, if sufficient recess is provided following land use practices such as selective logging and chena cultivation in spatial and temporal contexts.

Chapter 3 - Being a desert state, the climate of Rajasthan is unique characterized by low rainfall, dryness and extreme temperature. Fauna of this region are specific, adapted to this hostile environment. This unique diversity of species makes Rajasthan a regional priority for conservation. Recently the exclusive fauna of this region is facing many unprecedented threats. Nowadays, the most serious threat is climate change. Oil exploration, overgrazing, mining of minerals and sandstones, habitat alteration and destruction caused by Indira Gandhi NaharPariyojna canals, etc. are the other prevailing threats. Current knowledge on the major threats to fauna of Rajasthan is briefly reviewed here. According to World Wildlife Fund India, this area has lost half of its wildlife in past forty years. So studies on threats to fauna of Rajasthan will assume increasing relevance as government sectors seek to protect and conserve declining biodiversity. More interdisciplinary studies are required to warn about unnoticed threats to biodiversity of Rajasthan.

Chapter 4 - The Tamil people living in southern India and neighbouring regions, as well as in Diaspora (about 80 million), claim a unique pride in their Dravidian identity, their long history and the classic nature of their language. In post-Independence India (since 1947) they have had to struggle to safeguard their identity from many challenges. The Dravidian movement with its powerful proponents initiated a Tamil renaissance with their provoking speeches and writings, and cinema has played a major role. Consequently, the leading section of the movement captured political power in 1967, and one of their first activities was to change the name from Madras State (formerly Presidency) to Tamil Nadu (Tamil country). M. Karunanidhi (1924-2018), an avowed atheist and rationalist, popularly referred to as Kalaignar (artist), five times chief minister, an electrifying orator, with many honours including a doctorate, penned more than 70 film stories and dialogues (as well as 100 books), of which 39 hit the silver screen. The authors have made a selection of unequal length depending on the focus of the story and length of the narrative in order to study their communicative power and to undertake a content analysis. A school dropout from an ordinary family, an activist from a young age, his sacrificial determination to fight against religious superstition and caste discrimination and for the protection of the pure Tamil language against the imposition of the northern Indian Hindi was phenomenal. The introduction of motion pictures (1940s) gave him a unique opportunity to propagate the ideas of justice and equality with reference to myths and folklore. Films proved to be one of the cheapest means of entertainment, education and conscientization of the masses, most of whom became supporters and members of his political party. Some of his film stories have been most effective to make people think and get organised with awareness of the need of political power for effecting change. However, in due course, unexpectedly, his road towards the peak became rough with two rivals who came to fame by the same medium of cinema as actors, with a similar overarching appeal to the masses. Perhaps his life ended with an anti-climax when there was a legal battle to secure a burial place in a prime spot along with that of his mentor and two opponents. Does it have to do with the pedagogic nuances of cinema or using it with a hidden agenda of dominant forces which employed the tested tactics of countering and confusing ideologies? Who influenced the people more, the playwrights behind the screens or the actors with strategic titles and roles on them? After presenting a short introduction to the historical and cultural background the authors narrate in detail Karunanidhi's most decisive film *Parasakthi* (Supreme Power, 1952). It challenges certain cultural stereotypes (the sexual exploitation of women), social discrimination, corruption, religious superstition and so on. After an interlude of new challenges, *Neethikkuth Thandanai* (Punishment for Justice, 1987) narrates a fight against domination, injustice, corruption of the police and judiciary involving beastly rapes and gruesome revenges. *Puthiya Parasakthi* (New Supreme Power, 1996) corresponds to the film of 1952 but with greater violence. In between, a number of films are mentioned in passing which deal with certain aspects of social reality and their pedagogic power. The final part of the article, ‘Highlights and Questions’, has ten points of irony, frustration and depression.

Chapter 5 - This chapter examines the effects of socio-economic determinants on child malnutrition in Uttar Pradesh- the most populous and malnourished state of India. Using data from NFHS-4, this study highlights district-wise variations in nutritional status of children less than five years in the state. The percentage of stunted, wasted and underweight have been taken as dependent variables, while main explanatory variables of child malnutrition were female education, maternal malnutrition (BMI), breastfeeding practices, children’s place of residence, father’s education, wealth index of household, family planning, sex of the child, sex of the household and age cohort of mother. Logistic regression models have been applied in order to observe the association between explanatory variables and malnutrition. The findings revealed a high incidence of less stunted, wasted and underweight infants in case of an educated parents, particularly educational level of mother. Wealth index representing the possession of assets by the household found to be an important indicator in the sense that children born to households with richer and richest quintiles showing better-nourished children aged under-5. The modern method of using contraceptive, breastfeeding practices, maternal malnutrition were also important indicators influencing the rate of malnutrition in children. The study suggested that imparting education to females, breastfeeding practices, economic status and contraceptive use can have an important role in reducing nutritional status of children.

In: India
Editors: J. Lessard and O. Rivière
ISBN: 978-1-53618-413-6

Chapter 1

PHYSICAL AND DEMOGRAPHIC STUDY OF UTTARAKHAND

B. R. Pant*[1,*]*, PhD, R. Chand*[2]*, PhD and B. S. Mehta*[3]*, PhD

[1]Professor of Geography, M. B. Government P. G. College Haldwani, Uttarakhand, India

[2]Professor of Geography, Kumaun University Campus, Nainital, Uttarakhand, India

[3]Head, MIS/GIS, J. B. G. V. S., Viveknagar, Akurdi, Pune, Maharastra, India

ABSTRACT

Historically, Uttarakhand is considered as densely forested area, place of God and Goddesses and the home of wild animals and also got recognition as a confluence of various cultural groups came from Central Asia in the north, Nepal in the east and Himachal in the west and the Gangetic Plain in the south. Geographically it extends from the Tons - feeder of Yamuna River in the west to the Kali or Sharda in the east. Its

---

* Corresponding Author's E-mail: brpant123@rediffmail.com.

northern boundary is demarcated by Indo - Tibet water parting ridge separating it from Tibet (China) and southern boundary corresponds almost with the southern limit of Tarai belt separating it from Uttar Pradesh. Situated in the Himalayan Region, all the districts except Hardwar and U. S. Nagar of the Plain, Tarai and Bhabar region are mountainous tracts. Historically, the region under the present study has been divided into two divisions- Garhwal (Kedarkhand) and Kumaun (Manaskhand) – presently Garhwal includes the districts of Hardwar, Dehradun, Uttarkashi, Tehri, Rudraprayag, Chamoli and Pauri, and Kumaun spread over the districts of Almora, Bageshwar, Pithoragarh, Champawat, Nainital and U. S. Nagar. About 85% of the land area of the region is mountainous and remaining fall in the almost plain area- Bhabar, duns, Tarai and river Ganga plain. Lying between $28^0$44'to $31^0$25' north latitudes and $77^0$45' to $81^0$1' east longitudes, it encompasses an area of 53483 km^2 and supports 10086292 people in 2011. Due to the unfavorable physical conditions only 49% population is made their habitat in the mountainous region of the Uttarakhand state. Out of the total population 69.8% (7036954 persons) are residing in rural villages and the remaining 30.2% (3049338 persons) people are registered as urban population. Out of the total population 18.8% and 2.89% population belongs to scheduled caste and scheduled tribe population respectively while in rural Uttarakhand these categories have respectively 21.3% and 3.76% population. The average sex ratio in Uttarakhand is 963 females per thousand males. It is noticed that the sex ratio among children (0-6 years), scheduled caste and the scheduled tribe is less than the total population in these respective categories. The average literacy rate in Uttarakhand is 78.8%. The average growth rate of the region is 18.8% which is more than the country average.

The main aim of this study is to highlight the physical environmental conditions such as relief, structure, drainage, climatic conditions, forest types and influence of these aspects on demographic structure such as population distribution, growth, sex ratio, literacy, etc. of Uttarakhand. The present study is based on secondary data released by the Census of India, 2011 and literature available in the public domain. The findings of this study will be helpful to make a feasible plan for the all-round sustainable development of Uttarakhand.

Keywords: physiographic, demographic, migration, human resource, growth, scheduled castes, scheduled tribe, sustainable development

PHYSIOGRAPHIC/GEOGRAPHIC REGIONS

Uttarakhand comprises all the five latitudinal Physiographic and geological zones-the Tarai-Bhabar, the Siwalik with duns, the Lesser Himalaya, the Great Himalaya and the Trans Himalaya (Figure 2). These latitudinal divisions are separated from one another by thrusts/faults such as Himalayan Frontal Fault (HFF) exist between the Upper Siwalik and Bhabar, Main Boundary Thrust (MBT) lies between the Lesser Himalaya and Lower Siwalik, Main Central Thrust (MCT) separated the Lesser and the Great Himalaya and Trans Himadri Thrust (THT) lies between the Great and the Trans Himalaya (Pant, 1995 and 1998) The northern fringe of Indo-Gangetic plains, waterlogged in older times is known as Tarai. Due to the availability of fertile soil, hot climate coupled with abundant water, this 10-25 km. wide, once densely forested Tarai belt has been the homeland of Tharus, Buksas, and Gujjars, and is largely reclaimed for agriculture. Due to the introduction of new agrarian techniques by the enterprising hardworking refugees from Pakistan and Bangladesh, retired army personnel and freedom fighters of Uttarakhand, the Tarai today is one of the most developed parts of the country. In the Tarai, Bhabar, Dun and adjoining Siwalik belts, there are about 13% of rural settlements and 60% of urban centers including Dehradun, Rishikesh, Kotdwar Kalagarh, Ramnagar, Kashipur, Haldwani-cum-Kathgodam, Tanakpur, Khatima, Rudrapur, Sitarganj, Gadarpur, Kichha, Bajpur, Jaspur, Pantnagar, (Nagla), etc. There are several multipurpose reservoirs like Nanaksagar, Bahgul, Haripura, Tumeria and Kalagarh in the Tarai and Bhabar regions of Uttarakhand.

With an altitude varying from 700 to 1200 m, the Siwalik, like the Bhabar region, is sparsely inhabited because of thick vegetal cover and water scarcity. The gravel filled, almost plain latitudinal areas in the Siwalik domain are known as Duns. There are five duns namely, Dehradun between Yamuna and Ganga, Chaukhamba and Kothari duns between Ganga and Western Ramganga, Patlidun between Western Ramganga and Kosi, and Kotadun between Kosi and Baur. Course gravels in which all streams disappear form a 10-20 km wide belt between the Siwalik and

Tarai known as Bhabar. Consequently, the belt faces an acute problem of water scarcity. The 70 km wide Lesser Himalaya having relatively milder / stable terrain is situated north of the Siwalik Hills. This is the densely populated belt of Uttarakhand. The urban centers (37%) of Dwarahat, Ranikhet, Almora, Bageshwar, Dharchula, Pithoragarh, Lohaghat, Champawat, Bhimtal, Bhawali, Nainital, Lansdown, Narendranagar, Pauri, Srinagar, Tehri, Uttarkashi, Chamba, Mussoorie, Gopeshwar, Joshimath, Karnaprayag, Rudraprayag, Barkot, Purola, etc. are located in the Lesser Himalaya.

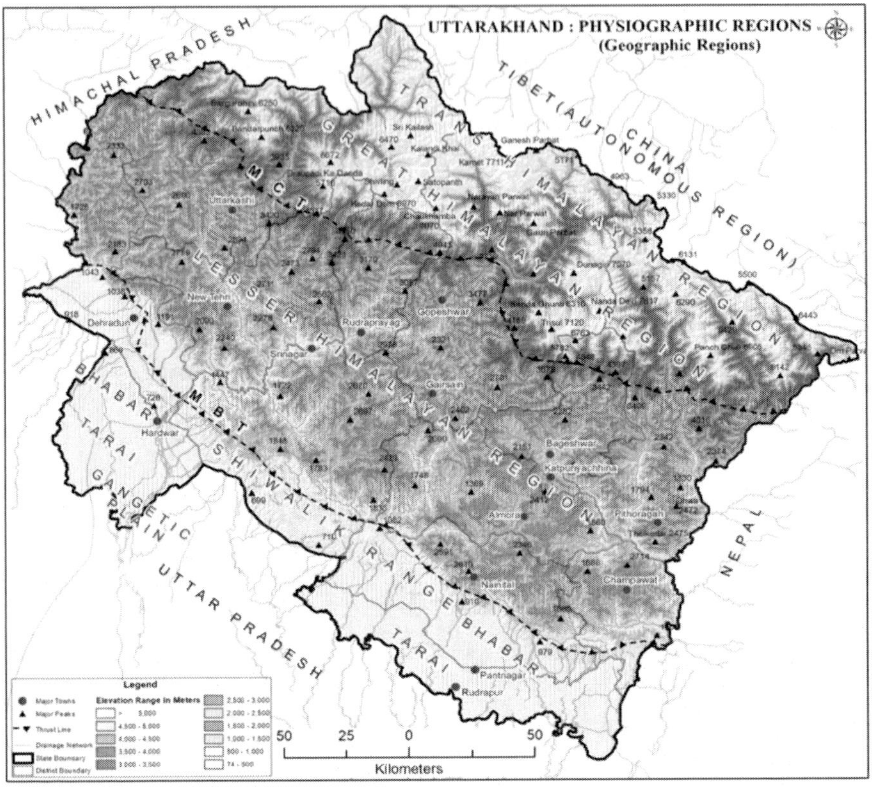

Figure 2. Physiographic/Geographic Regions of Uttarakhand.

There are about 81% of the rural settlements in the Lesser Himalaya. The Great Himalayan region in the north is a belt of very high peaks from 3000 m to more than 7000 m in altitude and of glaciers. It ranges in width from 30 to 50 km. The important peaks include Nandadevi, Panchchuli, Trishul, Kedarnath, Chaukhamba, Badrinath, Bandarpunch and Kamet. Among the glacier Pindari, Milam, Gangotri and Khatling are the important ones. The harsh and rigorous climatic conditions together with extremely rugged topography are perhaps responsible for scanty habitation. Consequently, there are less than 70 seasonal settlements in the river valleys and bugyals (alpine pasture).

The Trans-Himalaya belt is about 40 km wide. This is a rain shadow zone, as southeastern monsoon clouds cannot cross the Great Himalaya. The belt is scantily populated, there being just 20-30 temporary seasonal villages of the Bhotia scheduled tribe.

Geological Setup

From the geological point of view, the Uttarakhand or Kumaun Lesser Himalaya is divided into five litho tectonic units. These units are demarcated by several thrusts and faults. Most of these faults/thrusts are active. The brief description of the geology/structure is based on Valdiya (1980a 1980b and 1988).

The foothill belt of the Lesser Himalaya is made of sandstone, mudstone and subordinate conglomerates of the Middle Miocene to Early Pleistocene age. The Lower Siwalik consists of fined grained calcareous sub greywacke and sub lithogenic alternating with dominant horizons of maroon mudstone and rare Para conglomerates while the Middle Siwalik consists of predominant coarse-grained sub lithogenic and subordinate Para conglomerate. The Upper Siwalik sediments comprising boulder conglomerate with maroon mudstone and sandstone. Himalayan Front Fault (HFF) separates Siwalik from the Bhabar. The gravel filled synclinal valleys are known as Duns-Dehra, Chaukhamba, Kothari, Patli and Kota. It includes two litho-tectonic units-(1) Tarai and Bhabar, and (2) Siwalik.

Lying between the MBT in the south and MCT in the north, the Lesser Himalayan terrain is made of three litho-tectonic units:

1) The autochthonous succession of the Precambrian strata forming the larger part of the inner belt.
2) Late Precambrian to early Paleozoic sedimentary rocks formation forming the Krol Nappe includes slates, phyllites sandstone, and limestone.
3) The Precambrian metamorphic injected was a granite body.

In summing up there are seven main lithological groups namely-Krol Succession, Ramgarh group, Almora Group, Damtha-Tejam group, Bhatwari-Barkot formation, Berinag Formation, and Munsyari.

The Great Himalaya is made of high-grade metamorphic- the Vaikrita group of Precambrian age and intruded extensively by mid-Tertiary granite. However, Jhingran referred it to the Central Crystalline Zone.

Trans Himadri Thrust-separates the Trans Himalaya from the Great Himalaya. The rock formations are shattered mylonolized or otherwise complexly folded.

Drainage Network/River Basins

Uttarakhand is mainly drained by the Ganga and the Kali systems. Broadly there are five river basins or watersheds, namely, Yamuna and Tons, Bhagirathi and Alaknanda, Ramganga (East) and Saryu, Kali and Gori, Ramganga (West) and Kosi and Gaula and Others (Figure 3).

The Yamuna originates from the Yamunotri hot spring and Saptrishikund on the southwestern slope of the Bandarpunch ridge and flows through the district of Uttarkashi and Dehradun. The Tons is the largest river in respect to the volume of water flowing. Bhagirathi and Alaknanda join at Devprayag to make the Ganga River. The Bhilangana is the main tributary of the Bhagirathi that rises from Khatling glacier. Western Dhauli originates at Untadhura and Niti passes on the Indo-Tibet water divide and joins the Vishnuganga at Vishnuprayag. Saraswati, emanating near Mana Pass, makes the headwaters of the Alaknanda. Pindar (originating from Pindari glacier) and Mandakini (rising in the Kedarnath ridge) are the main tributaries of the Alaknanda. Nayar joining the Ganga from the west at Byasighat drains the Dudhatoli – Pauri region.

Kali is the largest river in Kumaun both in respect of the volume of water as well as the extent of the catchment area. It is also the largest river in Uttarakhand in respect of length. The Kali originates at Kalapani near the Lipulekh pass and is fed by Kuti, Eastern Dhauli, Gori, Saryu, Lohawati, and Ladhiya rivers. Gori emanates from the Milam glacier and Eastern Ramganga from Namik glacier. Eastern Ramganga and Panar are the principal tributaries of the Saryu. Saryu joins the Kali at Pancheswar. The Kali is known as Sharda River at Baramedeo mandi.

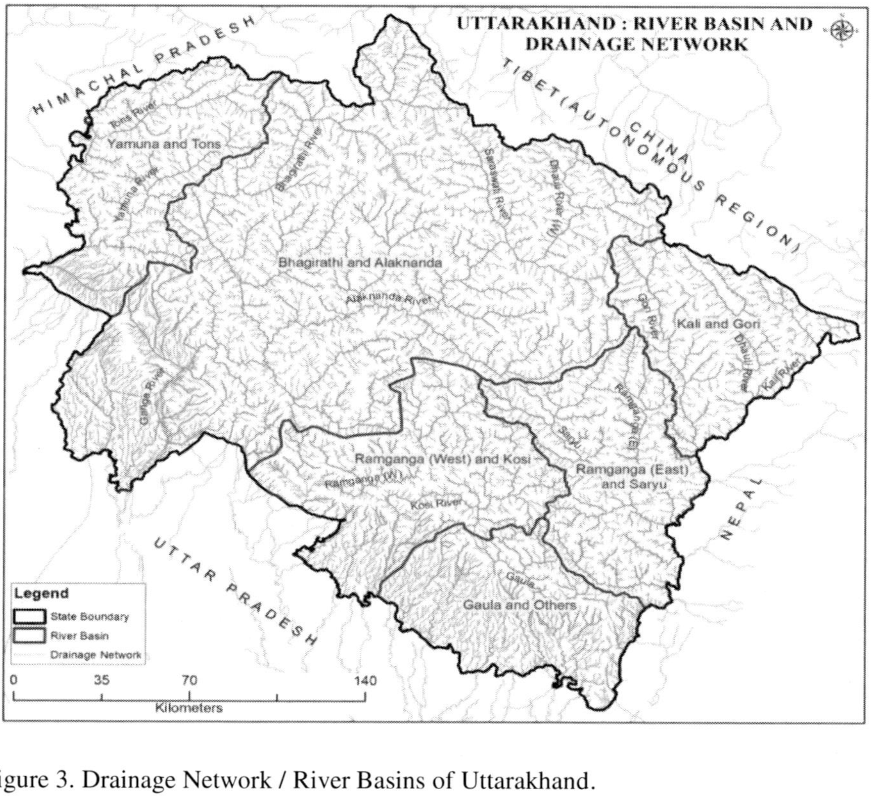

Figure 3. Drainage Network / River Basins of Uttarakhand.

The Western Ramganga rises from the northeast slope of Dudhatoli ridge in the Pauri district. Bino the headwater of Western Ramganga takes its rise from the Binsar spring located on the southern slope of the Dudhatoli range (Baramdhungi). It joins the Masangaddi and Western Ramnagar at Deghat and Kedar, respectively. Rising from the Bhatkot ridge, Gagas joins the W. Ramganga at Bhikiasain. The Kosi River rises on the southern slope of Bhatkot- Kausani range and leaves Kumaun near Sarkara eventually to join the Western Ramganga.

Minor rivers like Song, Khoh, Rispana, Dhela, Dabka, Baur, Bhakhara, Gaula, Nandhaur and Deoha; drain the outer belt of the Lesser Himalayan domain. Due to the porous nature of Bhabar areas these rivulets became dry and reappear in the Tarai region where these are checked in the form of artificial reservoirs for multipurpose.

Climatic Conditions

Among all the environmental influences that control human activities, the climate seems to be the most effective constraint particularly in determining the lifestyle in a region.

Diversified topography of the region has offered different climatic features. Rainfall and Temperature is an important factor which determines the cultivation and human activities. The altitudinal difference, slope aspect and forest cover of the area create complex climatic conditions that change frequently from location to location. The climate in the river valley is quite in contrast to the climate on the higher ridges. The Tarai-Bhabar and low-lying valleys below 800 m of elevation have hot and humid with a subtropical climate. Between 800-1200m elevation, it is warm-temperate, from 1200 to 2000m it is cold temperate, between 2400-3000m, elevation cold climate; 3000-4200m of elevation alpine climate; and above 4200m elevation glacial climate (Figure 4).

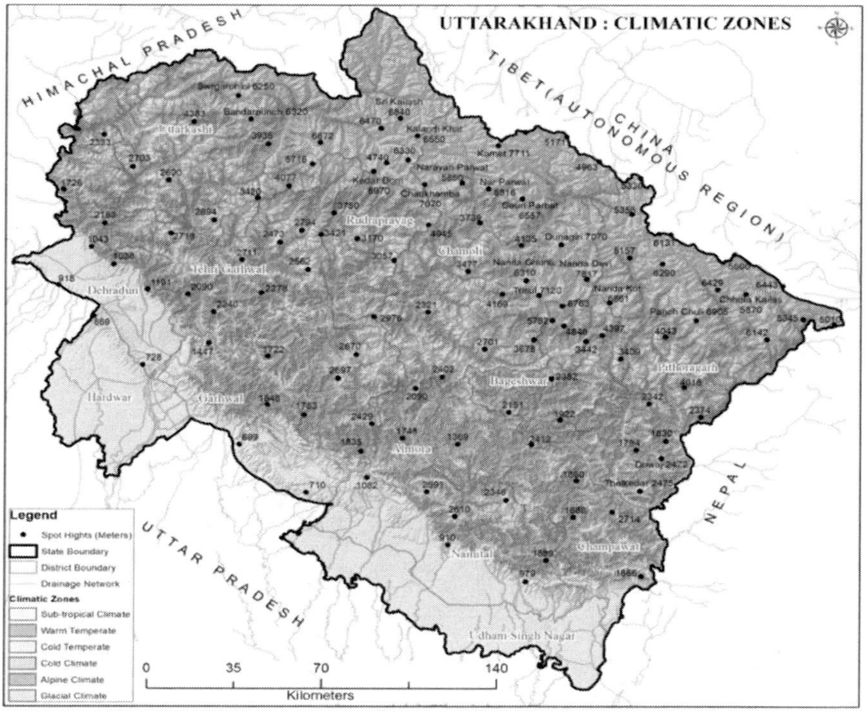

Figure 4. Climatic Zones of Uttarakhand.

In a large measure, the climatic distribution in the hills can best be explained by one of the fundamental principles of vertical distribution which is largely a result of the vertical distribution of temperature falls by as much as 6o C but large variations may occur depending upon such factors like location, time of the and the season (Joshi et al., 1983). The temperature decreases from the valley upwards and finally reaches the freezing point in the northern Trans Himalayan tract. The northern and western papers of the region have low temperature while in the central and northern part of the region the temperature increase along the valleys. Due to this character of temperature, the cropping season and duration of harvesting and agricultural practices differ from region to region. The mean temperature in January is about 210C in the sub mountain belt of Tarai, Bhabar and Duns while 80C in the inner valleys. The maximum temperature register is around 39.70C in Tarai Bhabar (Pantnagar), 41.50C in Dun valley (Dehradun) and 34.620C in the Lesser Himalayan domain (Tehri) in the year 2004 and minimum temperature, 3.20C in Tarai Bhabar and Dun valley and 0.60C in the lesser Himalayas respectively. A similar trend has been recorded in temperature in the years 2005 and 2006.

There is a general tendency of a decrease in the annual rainfall from east to the west though it is not well marked. The mean monthly rainfall of the Uttarakhand region is about above 630 mm. The maximum rainfall occurs at two points, where a general rise in elevation fakes place at the foot of the utmost range of hills, and the foot of the snow. Most of the areas are rainfall data is not available lies along the high altitude zone of Himalaya, where precipitation always occurs in the form of snow. About 80% of the total rainfall occurs in the monsoon months alone. The average annual rainfall is 150cm in the Tarai-Bhabar and Dun belts, while the Nainital hills receive the highest i.e., 250 cm. then decreasing northwards in the inner Lesser Himalaya where the average annual rainfall is only 90 cm (Pant, 1994). The Trans Himalayan region falls in the rain-shadow zone because the southeast monsoon clouds cannot cross the Great Himalaya. The rainfall record in the year 2004 is around 1415.3 mm in Tarai - Bhabar (Pantnagar), 2026.8 mm in Dun valley (Dehradun) and 1205.6 mm in the Lesser Himalayan region (Tehri).

VEGETATION/FOREST COVER

The emergence of Himalayan ranges that took place in several phases brought about a corresponding change not only in the climatic profile along with the altitudinal gradient but also in the soil factors. These changes influenced the immigration of plant species from far off regions, from east and west, their establishment in the ecosystems speciation and extinction during various geological ages. The trend of colonization and the formation of stabilized communities followed by speciation in the Himalaya continued even in the present age. Geographically and ecologically, Uttarakhand is therefore, one of the most complex areas in the Indian sub-continent with marked east-west and south-north transitions.

Based on the interpretation of satellite data about Oct.-Dec.2015, the forest cover in the state is 24,295 sq km which is 45.43% of the state's geographical area. In terms of forest canopy density classes, the state has 4969sq km under very dense forest, 12884 sq km under moderately dense forest and 6442 sq km under open forest. The State has reported a recorded forest area of 38,000 sq km which is 71.05% of its geographical area. The Reserved, Protected and Un-classed Forests are 69.86, 26.01 and 4.13% respectively of the recorded forest area. However as the digitized boundary of recorded forest area from the State covers only 28,186 sq km, the analysis of forest cover within this area is depicted in Table 1 and Figure 5.

Table 1. Forest cover in Uttarakhand, 2017
(area in sq.km)

A	**Forest Cover within Recorded Forest Area**	
01	Very Dense Forest	4172
02	Moderate Dense Forest	9344
03	Open Forest	3264
	Total (1+2+3)	16780
B	Forest Cover outside Recorded Forest Area	
04	Very Dense Forest	797
05	Moderate Dense Forest	3540
06	Open Forest	3178
	Total(4+5+6)	7515
	Total Forest Cover (A+B)	24295
C	Tree Cover	767
	Total Forest and Tree Cover (A+B+C)	25062
	Of State's Geographical Area	46.86%
Of India's Forest and Tree Cover		3.12%
Per Capita Forest and Tree Cover		0.25 ha

Source: India States of Forest Report (ISFR), 2017 page- 309.

The total carbon stock of forest in the state is 284.664 million tonnes (1043.768 million tonnes of Co2 equivalent) which is 4.02% of total forest carbon of the country (India States of Forest Report, 2017 page- 311.).

Forest Cover/Types

Uttarakhand state lies mainly in the Himalayan range, the climate and vegetation vary greatly with elevation from glaciers at the highest elevation to the subtropical forest at the lower elevation. The high altitudinal zones are covered by snow and bare rocks while lower regions such as Tarai, Bhabar, Duns and lower river valleys are covered by dense forests or vegetation. Owing to the altitudinal, climatic, structural, slope aspects, etc. variations, Uttarakhand state has been divided into thirty-four forest types (Table 2).

Table 2 shows that the maximum 26.07% area of the total forest area of Uttarakhand has been covered by the Himalayan Chir Pine forest followed by the Banj Oak Forest with 20.23% of total forest area. Moist Siwalik Sal Forest stands in the third rank with a 12.97% forest area.

Source: India States of Forest Report (ISFR), 2017).

Figure 5. Forest Cover in Uttarakhand, 2017.

Table 2. Forest types in Uttarakhand

Sl. No.	Forest Type	Area in sq. km.	Area in %
01	Moist Siwalik Sal Forest	3212.21	12.97
02	Moist Tarai Sal Forest	542.2	2.19
03	West Gangetic Moist Mixed Deciduous Forest	879.9	3.55
04	Dry Siwalik Sal Forest	364.4	1.47
05	Dry Plain Sal Forest	57.05	0.23
06	Northern Dry Mixed Deciduous Forest	697.88	2.82
07	Khair- Sissu Forest	242.35	0.98
08	Dry Deciduous Scrub	216.98	0.88
09	Lower or Siwalik Chir pine Forest	449.41	1.81
10	Upper or Himalayan Chir pine Forest	6454.61	26.07
11	Himalayan Subtropical Scrub	165.19	0.67
12	Sub tropical Euphorbia Scrub	42.04	0.17
13	Banj Oak Forest	5009.21	20.23
14	Moru Oak Forest	978.92	3.95
15	Moist Deodar Forest	486.36	1.96
16	Western Mixed Coniferous Forest Spruce, Blue Pine, Silver Fir	541.95	2.19
17	Moist Temperate Deciduous Forest	264.29	1.07
18	Oak Scrub	76.5	0.31
19	Himalayan Temperate Secondary Scrub	25.02	0.1
20	Kharsu Oak Forest	244.59	0.99
21	West Himalayan Upper Oak / Fir Forest	1131.34	4.57
22	Moist Temperate Deciduous Forest	317.71	1.28
23	Alder Forest	10.16	0.04
24	Low Level Blue Pine Forest	3.99	0.02
25	Dry Deodar Forest	362.96	1.46
26	Hippophae / Myricaria Scrub	85.8	0.35
27	West Himalayan Sub- Alpine Fir Forest	194.49	0.78
28	West Himalayan Sub- Alpine Birch / Fir Forest	611.84	2.47
29	Hippophae / Myricaria Brakes	19.23	0.08
30	Deciduous Sub- Alpine Scrub	211.94	0.86
31	Birch / *Rhododendron* Scrub Forest	138.44	0.56
32	Dwarf *Rhododendron* Scrub	32.36	0.13
33	Dry Alpine Scrub	5.93	0.02
34	Dwarf *Juniper* Scrub	33.57	0.14
35	Plantation / TOF	651.18	2.63
	Total	24762.00	100.00

Source: Atlas Forest Types of India, 2011 Quoted by ISFR, 2015: 253-254p.

SOILS

A complete soil survey of the Uttarakhand Himalaya has not been carried out so far. Based on the available data, it becomes difficult to determine the remarkable characteristics of the soil. Near Dehradun of the foothill is show the soils are loam to clay-loam in texture and acidic. In the Ranikhet area four broad soil groups have been identified – red loam, brown forest soil, podzolic soil and meadow soil (Ghildyal, 1981). Based on altitude, structure, landscape, slope aspect, gradient, vegetation, drainage and fieldwork, the soil of Kotadun, a southwestern part of Kumaun Himalaya is classified into seven groups – loam soil, conglomeratic sandy soil, stony hilly soil, crushed salty soil, hill sandy soil, podzolic soil and rocky thin regolith soil (Pant, 1988 and 1994). The soil composition in the Himalayan belt indicates that two different and distinct types of soils are present in these areas, immature soils, and natural soils. It is known as the cumulative interaction of climate, vegetation and topography work together on the weathered product of the disintegration of rocks to make a type of soil. In the lofty mountainous areas of the study region, it is fairly evident, that soils are not adequately mature due to only moderate chemical weathering, a low mass movement, fluvial erosion and slow wind action as coupled with very low temperature. Sedentary soils have low maturity and skeletal presence despite their grater erosion of slopes. The snow-clad areas do not allow an adequate chance for the rocks to disintegrate and undergo all chemical processes to make mature soils. The valleys, where soil from various slopes are gathered by the action of dashing water, bear the drift soils of varying composition as distinct from the sedentary soils. The valleys have alluvial deposits and them afford a good crop-growth. Land use strategy, thus, may include the sub-regional soil demarcation and appropriate crop-combinations for them. The sedentary soil brings adequate humus content with them as this flow with water currents on forested slopes.

Therefore, the alluvial soil collected at valleys has fineness, hummus, phosphorus gained from bones on way, and the admixture of various plant nutrients. Land use is governed by the relief and nature of the terrain in the altitudes. The situation of the underground water table, stage of maturity of soil, air drainage and absence of frost has bearing on the potentiality of crops in the study region. Towards the 'Duns', Bhabar and Tarai, the texture of the soil is good for cropping. The soils of Bino, Western Ramganga. Kosi, Saryu and Gomti river valleys are superb. Keeping in mind the structural characteristics of the region, the soils may be divided into eight groups, namely, Bhabar soils, Tarai soils, Older Alluvial soils, Younger Alluvial soils, Skeletal soils, Sub mountain soils, Mountain Meadow soils and Brown Red and Yellow soils (Figure 6).

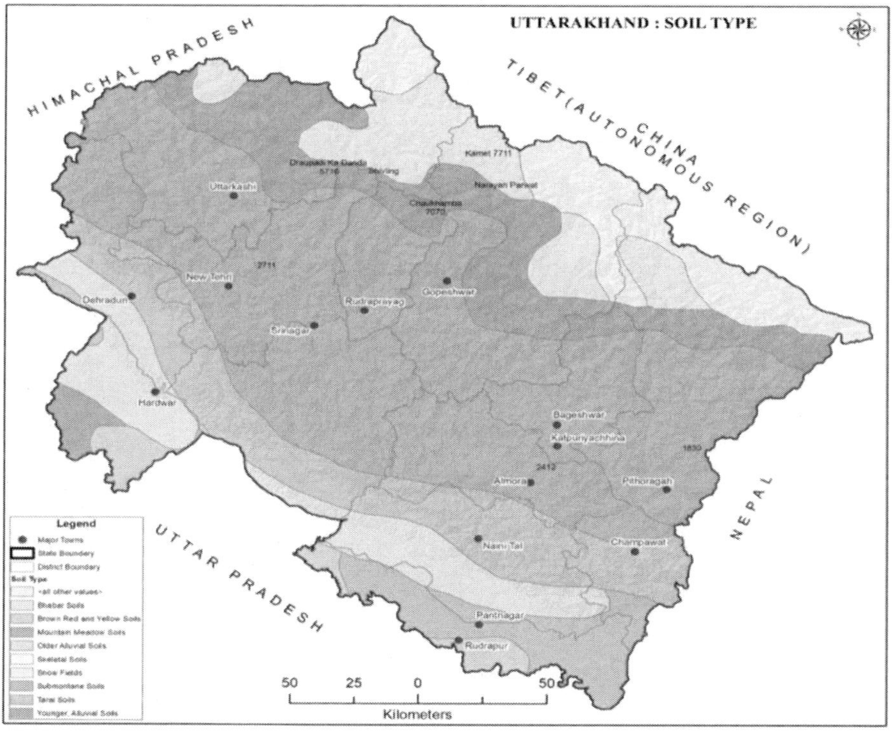

Figure 6. Major Soils of Uttarakhand.

DEMOGRAPHIC CHARACTERISTICS

Uttarakhand region presents very distinct demographic characteristics determined by the local relief and the harsh climate together with a host of other physical as well as cultural factors. There are areas with no population in the Trans Himalaya to densely populated areas along the foothills. The distribution pattern of the population not only differs from the rest of the country but greatly varies within the states and districts belonging to one part to another part of the region. As one proceeds from the southern Tarai –Bhabar tract to Siwalik, Lesser Himalaya; and Greater Himalaya and Trans Himalayan region of the state, the population decreases markedly by the topography and environmental conditions (Chand et al., 2016). The slope gradient primarily controls the clustering of population and now the infrastructural development has altered and shifted the population nucleus from their original locations to the places receiving developmental inputs (Chand, 2013 and 2017). About two-thirds population of the state is concentrated in 4 districts falling mainly in the plains areas while the remaining 40% population is distributed among 9 hill districts. According to Census 2011, there is an increase of about 18.81% (1.5 million) from 2001. Notable here is the fact that the increase in the plains area district is about 31%, while it has been just 3% in hill areas. Almora (-1.64%) and Pauri (-1.41%) districts recorded a decrease in total population. The district of Dehradun (32.33%), Nainital (25.1%) and Udham Singh Nagar (25.1%) has shown high growth. These reflect the trend of migration towards the districts in plains having big cities and centers of industrial and economic development and better opportunities for living.

Population Growth, Distribution, and Density

Uttarakhand is one of the most thinly and unevenly populated regions in the country. However, when compared within the Himalayan states, Uttarakhand ranks next to Jammu & Kashmir with 20.87% of the total

Indian Himalayan states. The distributional pattern and growth of the population of Uttarakhand are quite distinctive from those of the other Himalayan states. The trends and patterns of the human concentration do reflect considerable changes in both in Spatio-temporal perspectives. The population estimates of the first census conducted in 1872 suggested a total of 11, 77, 329 persons in Uttarakhand excluding Hardwar district. The annual growth of the population between 1872 and 1901 in Uttarakhand was estimated somewhere around 2%. If we consider 1901 as a base year, the population of Uttarakhand grew by 405% during these 110 years up to 2011% (Pant. 1995, Pant and Chand, 2014).

Table 3. Population growth/change in Uttarakhand since 1901 (in%) Census Year

District/ Region	1901-1911	1911-1921	1921-1931	1931-1941	1941-1951	1951-1961	1961-1971	1971-1981	1981-91	1991-2001	2001-2011
Uttarkashi	11.9	5.8	9.8	13.7	3.7	15.8	20.3	29.2	25.5	23.1	11.9
Chamoli	11.7	1.2	10.0	12.8	6.2	19.5	17.0	24.1	22.6	13.9	5.7
Rudraprayag	11.7	2.4	10.0	13.0	5.5	13.3	13.1	25.1	18.1	13.4	6.5
Tehri	11.9	5.8	9.8	13.7	3.7	13.0	14.3	24.7	16.5	16.2	2.3
Dehradun	15.3	3.6	8.5	15.6	36.1	18.6	34.6	31.9	34.7	25.0	32.3
Pauri	11.7	1.2	10.0	12.8	6.2	14.0	14.5	15.5	8.6	3.9	-1.4
Pithoragarh	15.9	0.9	10.0	17.8	12.5	19.1	14.4	16.4	14.1	11.0	4.6
Bageshwar	15.9	0.9	10.0	17.8	12.5	14.3	24.0	19.6	14.8	9.3	5.2
Almora	15.9	0.9	10.0	17.8	12.5	13.6	15.2	15.8	8.9	3.7	-1.6
Champawat	13.3	-1.3	8.8	16.4	12.7	30.5	45.1	25.3	26.4	17.6	15.6
Nainital	-0.1	-14.4	0.2	5.3	14.9	37.6	23.1	38.1	30.2	32.7	25.1
U.S. Nagar	-0.2	-14.4	0.2	5.3	14.9	121.1	49.7	48.1	38.3	33.6	33.45
Hardwar	-5.6	-4.9	11.4	13.1	14.6	18.0	32.9	32.7	26.3	28.7	30.6
Uttarakhand	8.2	-1.2	8.7	13.6	12.7	22.6	24.4	27.4	23.1	20.4	18.8
India	5.7	-0.3	11.0	14.2	13.3	21.5	24.8	24.7	23.9	21.5	17.7

Source: Census of India, Primary Census Abstract, PCA0500 to 0513_2011_MDDS Uttarakhand.

A glance at retrospect (Table 3) reveals that Uttarakhand grew in the first decade of the twentieth century (1901-11) by 8.20% which was higher than the national average of 5.75%. The decadal population growth registered by the state before 1951 was less than 15% with a negative growth of -1.23% from 1911 to 1921. During these years from 1921 to

1951, the state was just behind the national growth rate. However, the growth was higher than 20% in each decade in the post-independence period till 2001. The highest growth rate of the twentieth century was recorded in the decade 1971-1981 by 27.45%. The post-independence period has also witnessed higher growth than the national average in Uttarakhand. The emphasis has been primarily laid here on the district level growth of the population from 1901 to 2011 (Figure 7). District wise population growth of Uttarakhand is shown in Table 3. The overall pattern suggests that there is a gradual decrease in the growth rate in all districts in each decade from 1981 to 2011 barring few exceptions. District Nainital is a lone district which has registered a slight increase in 1991 to 2001 than the previous decade of 1981-1991. Similarly, Dehradun has also recorded more growth in the decade 2001-2011 than the previous decade and has touched the highest growth rate from 1981 to 1991. The maximum growth of the decade 1981-1991 which was also the highest in all decades attained by district Udham Singh Nagar being more than 40%. The next highest in the later was district Dehradun and Champawat in the decade 1981-1991. During 1991-2001, Udham Singh Nagar is at the top closely followed by Nainital with more than 30% of decadal growth. These are all located in plain parts of the state. During 2001-2011, Udham Singh Nagar (33.45%) is so closely followed by Dehradun (32.33%) and Hardwar (30.63%) districts. These situations are the results of a heavy concentration of population in the plain areas of the state. On the other hand, all the hilly districts are below 20% in all decades with the exception of Uttarkashi which has also remarkably fallen to 11.89% growth in 2001-2011 and surprisingly that was the highest growth rate in 2001-2011. The current decade has finally displayed the negative growth for two hilly districts viz. Almora -1.64 and Pauri–1.41% (Table 3). Based on the growth trend so far revealed by the hill districts, it is assumed that all hill districts would be on the negative growth rate in the following decade of 2011-2021.

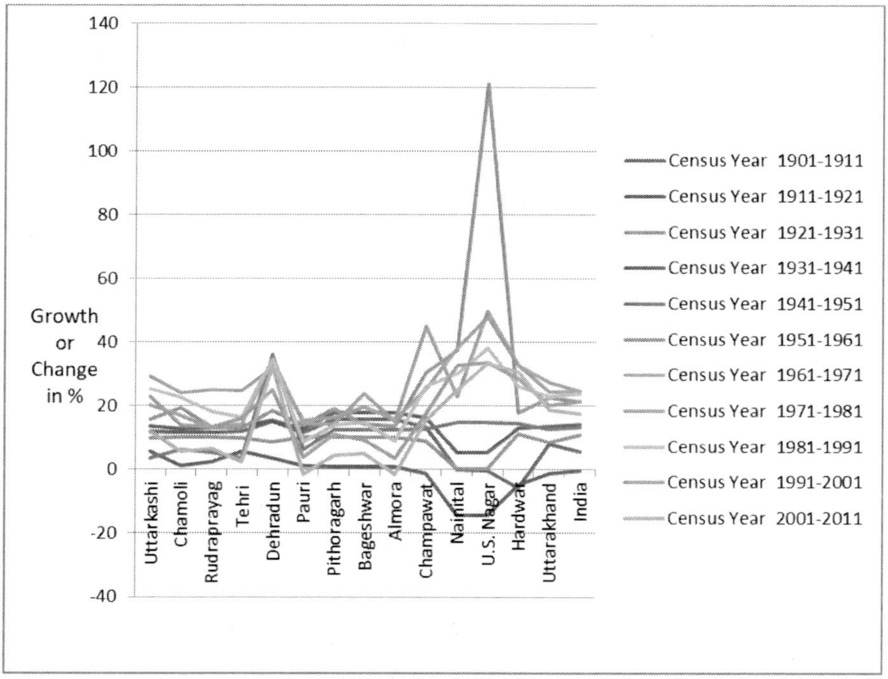

Figure 7. District wise Decadal Population Growth/ Change (%) in Uttarakhand.

As observed through the decadal growth rates, the district level data has shown a widening gap between the plain and the hill districts of the state. It is further confirmed by the population measured in terms of density per km^2. The average density of the population in Uttarakhand is 189 persons per km^2 in 2011 which was merely 37 persons per km^2 in 1901. In 2011, the highest density of population was measured in Hardwar (801) followed by Udham Singh Nagar (649) and Dehradun (549). The density is heavier in the southern plain parts of the state which lessens as we move northward in hilly areas. The lowest density of 41 persons per km^2 in 2011 is recorded in Uttarkashi with the next minimum of 49 persons per km^2 in Chamoli and 68 persons per km^2 in Pithoragarh. These are the areas of Greater and Trans Himalayan regions. Other hilly districts falling mainly within the Lesser Himalayan terrain have a moderate density between 122 to 198 persons per sq.km (Table 4). It is supported by the fact that during the decade 2001-2011, the maximum growth of population is attained by the same districts with the highest growth in Udham Singh Nagar (33.45%) followed closely by Dehradun (32.33%), Hardwar (30.63%) and Nainital (25.13%). On the other hand population growth in hilly districts during 2001-2011 was very low such that 2.35% in Tehri, 4.58% in Pithoragarh, 5.15% in Bageshwar, 5.74% in Chamoli, 6.53% in Rudraprayag and 11.89% in Uttarkashi. Even district Champawat recorded merely 15.63% as a sizeable part of it lies in plains. The most alarming situation was the negative growth recorded in Almora (-1.64%) and Pauri (-1.14%). There are only eight districts in the whole of the Himalayan region which have negative growth from 2001 to 2011(Pant and Chand, 2013 and Pant, 2013). The slowing down of the population in hilly districts had started from 1981 and it remained unnoticed till 2011 until it turned up in the negative whereas the increasing trend in the case of plains areas started especially after 1991.

Table 4. Some Silent Demographic Features of Uttarakhand, 2011

District/ Region	Area (Sq km)	Population	Density (P/Km^2)	Inhabited Villages	Towns	Development Blocks	Sub-Districts
Uttarkashi	8,016	3,30,086	41	694	3	6	6
Chamoli	8,030	3,91,605	49	1170	6	9	6
Rudraprayag	1,984	2,42,285	119	653	2	3	3
Tehri	3,642	6,18,931	169	1774	7	9	7
Dehradun	3,088	16,96,694	550	731	22	6	6
Pauri	5,329	6,87,271	129	3142	8	15	9
Pithoragarh	7,090	4,83,439	69	1572	3	8	6
Bageshwar	2,246	2,59,898	116	874	1	3	4
Almora	3,139	6,22,506	198	2184	5	11	10
Champawat	1,766	2,59,648	147	662	4	4	4
Nainital	4,251	9,54,605	225	1097	11	8	8
U. S. Nagar	2,542	16,48,902	648	674	19	7	7
Hardwar	2,360	18,90,422	817	518	24	6	3
Uttarakhand	53,483	1,00,86,292	189	15745	115	95	78

Source: Census of India, Primary Census Abstract, PCA0500 to 0513_2011_MDDS Uttarakhand.

The distribution of population and change 2001-2011 also confirmed that it has not been spatially uniform. Although the population has grown by 15, 96,943 from 84, 89, 349 in 2001 to 1, 00, 86, 292 in 2011. The population has grown only in the plain areas of the state already heavily populated. In 2011, about 52% population lives in Hardwar, Dehradun and Udham Singh Nagar alone and the rest 48% population lives in ten districts. These three districts also share the maximum population in the whole Indian Himalayan region. Hardwar is at the top (4.1%) followed by Dehradun (3.62%) and Udham Singh Nagar (3.51%) being the heavily populated districts in the Indian Himalaya in 2011 (Pant and Chand, 2013, Pant, 2013). A comparison of 2001 and 2011 data between plain and hill districts of the Uttarakhand state revealed that the population of plain Districts has grown from 4.7 million to 6.2 million with a net increase of 31%. On the other hand the hill districts have grown from 3.7 million to 3.9 million with a net increase of 3%. District Almora (-1.64%) and Pauri (-1.41%) have even shown negative growth.

The loss of population if viewed at sub- district (Tehsil) level presents more alarming situation. In the hills, out of 78 sub districts, 16 (20.06%) sub districts have registered decline in the population from 2001 to 2011. It is (-) 13.8% in Lansdown sub district of Pauri district. Followed by Chaubattakhal (-13.1%), Bhikiyasain (-11.8%), Satpuli (-11.4%), Dhoomakot(-10.9%), Tehri (-9.8%), Sult (-7.2%), Pauri (-7.1%), Jakhnidhar (-6.8%), Dwarahat (-6.1%), Chaukhutiya (-6.1%), Didihat(-4.6%), Yamkeshwar (-4.5%), Kanda (-3.0%), Ranikhet (-1.8%), and Thalisain (-1.2%) (Table 5). If the same trend is continues, 28 (35.9%) sub districts would further decline their population in 2021.These sub district are Chamoli (8.7%), Ukhimath (8.5%), Garud (8.5%), Nainital (8.1%), Jakholi (7.9%), Srinagar (7.8%), Dharchula (7.5%), Bhatwari (7.5%), Dunda (7.2%), Betalghat (6.6%), Almora (6.5%), Kalsi (5.9%), Pratapnagar (5.8%), Bageshwar (5.4%), Gairsain (5.1%), Kapkot (5.0%), Rudraprayag (3.8%), Kosya Kutauli (3.4%), Ghansali (2.4%), Gangolihat (2.2%), Bhanoli (2.2%), Karnaprayag (1.5%), Devprayag (1.4%), Pokhari (1.2%), Jainti (1.1%), Someshwar (0.9%), Tharali (0.8%) and Berinag (0.6%). Munsiari sub district which population has registered no change during 2001 to 2011 and it would further decline their population in 2021.

Table 5. Sub-districts by ranges of decadal (2001-2011) total growth in the Uttarakhand

Growth Zone/Region	Growth Ranges (%)	Uttarakhand		
		No	%	Sub Districts (Figures shown in brackets are in %)
Negative	(-)0.0-13.8	16	20.6	Thalisain (-1.2), Ranikhet (-1.8), Kanda (-3.0), Yamkeshwar (-4.5), Didihat (-4.6), Chaukhutiya (-6.1), Dwarahat (-6.1), Jakhnidhar (-6.8), Pauri (-7.1), Sult (-7.2), Tehri (-9.8), Dhoomakot (-10.9), Satpuli (-11.4), Bhikiyasain (-11.8), Chaubattakhal (-13.1) and Lansdowne (-13.8).
No Change	0.0	01	1.3	Munsiari
Low	0.01-10.0	28	35.9	Chamoli (8.7), Ukhimath (8.5), Garud (8.5), Nainital (8.1), Jakholi (7.9), Srinagar (7.8), Dharchula (7.5), Bhatwari (7.5), Dunda (7.2), Betalghat (6.6), Almora (6.5), Kalsi (5.9), Pratapnagar (5.8), Bageshwar (5.4), Gairsain (5.1), Kapkot (5.0), Rudraprayag (3.8), Kosya Kutauli (3.4), Ghansali (2.4), Gangolihat (2.2), Bhanoli (2.2), Karnaprayag (1.5), Devprayag (1.4), Pokhari (1.2), Jainti (1.1), Someshwar (0.9), Tharali (0.8) and Berinag (0.6).
Medium	10.01-20.0	15	19.2	Champawat (19.9), Poornagiri (18.9), Mori (18.8), Kaladhungi (18.4), Puraula (17.7), Pati (14.0), Chiniyalisaur (13.8), Narendranagar (13.6), Rajgarhi (13.3), Dhari (13.3), Chakrata (13.2), Dhanaulti (12.9), Pithoragarh (12.5), Tyuni (12.4) and Lohaghat (11.4).
High	20.01-30.0	9	11.5	Ramnagar (29.1), Roorkee (28.0), Jaspur (25.9), Laksar (24.8), Khatima (23.9), Kotdwara (23.1), Sitarganj (22.5), Lalkuan (21.2) and Joshimath (20.8).
Very High	Above 30.01	9	11.5	Kichha (49.4), Haldwani (42.4), Kashipur (40.2), Rishikesh (37.9), Hardwar (37.1), Vikasnagar (36.4), Dehradun (33.1), Bajpur (30.6) and Gadarpur (30.2).
	Total	78	100.0	----------------------------------

Source: Census of India, Primary Census Abstract, PCA0500 to 0513_2011_MDDS Uttarakhand.

On the positive side, there are 18 (23%) sub-districts with more than 20.01% growth, but most of them are in the plain area. These sub-districts are Kichha (49.4), Haldwani (42.4), Kashipur (40.2%), Rishikesh (37.9%),

Hardwar (37.1%), Vikasnagar (36.4%), Dehradun (33.1%), Bajpur (30.6%), Gadarpur (30.2%), Ramnagar (29.1%), Roorkee (28.0%), Jaspur (25.9%), Laksar (24.8%), Khatima (23.9%), Kotdwara (23.1%), Sitarganj (22.5%), Lalkuan (21.2%) and Joshimath (20.8%).

The sub-districts with having better growth may be characterized by better dynamics revealing their prospects for development. Not surprisingly, they all are comparatively in a developing situation even before the creation of the state. The sub-districts experiencing low or negative growth have usually shown, by nature a decaying condition that is static even after the state formation. Also, to comprehend the dynamics of the population in Uttarakhand, a micro-level analysis at the village level combining the aspects of aggregate percentage change and variability would help to assign different categories of change at different levels from villages to, blocks to the district level.

High change and high variability have been associated solely in foothill and plain areas at all levels. Those districts which are partially located both in hills and plains such as Nainital, Pauri and Champawat belong to the category of high change but moderate variability. There are two categories in the hill districts. Moderate change and moderate variability districts are newly created such as Bageshwar and Rudraprayag. However, the rest of the districts may be placed in the category of low change and low variability or nil change. They are in the static condition or slowly creeping on until they also join the category of negative growth like Almora and Pauri. The census figures as analyzed here signaling for a very high imbalance in the demographic behavior of the state.

Changing Sex Ratio

Demographic change on national, regional and local scales has important impacts on sex ratio and the social health of any population is explained by it. The earliest estimates available for the year 1872 noted 870 females per 1000 males in Uttarakhand. In the Garhwal region, however the females were much higher 1015 per 1000 males in 1871. On the other hand the females in Kumaun were numbered 893 in 1852, 891 in 1864 and 898 in 1872. In the year 1911, the lowest sex ratio was in Dehradun (733) and Nainital (796). The female ratio in the cities was even less in Pauri (375), Dehradun (574) and Nainital (668) in the same year (Pant and Chand, 2014).

Table 6. Sex ratio (females/1000 males) in Uttarakhand, 2001 &2011

Districts/Region	2001	2011		(0-6 age group) 2011	
	Total	Total	Urban	Total	Urban
Uttarkashi	959	958	838	916	794
Chamoli	1021	1019	767	889	829
Rudraprayag	1115	1114	697	905	803
Tehri	1049	1077	817	897	846
Pauri	1106	1103	917	904	860
Pithoragarh	1031	1020	913	816	724
Bageshwar	1106	1090	927	904	845
Almora	1145	1139	848	922	861
Dehradun	887	902	886	889	846
Hardwar	865	880	866	877	865
Champawat	1021	980	890	873	825
Nainital	906	934	912	902	892
U. S. Nagar	902	920	903	899	892
Uttarakhand	962	963	962	890	868

Source: Census of India, 2001and 2011.

There are several factors such that the male-dominated migratory practices are responsible for this lower sex ratio. The situation has improved gradually in the following census years. The sex ratio in Uttarakhand was 962 in 2001 which rose to 963 in 2011. The urban sex ratio also increased from 845 in 2001 to 884 in 2011 (Table 6). It is higher than the national average in both the census years from 933 in 2001 to 940 in 2011. The average sex ratio in the plain districts is 904 and the same in the hills is 1065 females per 1000 males in 2011. In seven hill districts, there are more females in Chamoli (1016 and 1019), Rudraprayag (1115 and 1114), Tehri (1049 and 1077), Pauri (1106 and 1103), Pithoragarh (1031 and 1020), Bageshwar (1106 and 1090) and Almora (1145 and 1139) both in 2001 and 2011. In Champawat the females have decreased from 1021 in 2001 to 980 in 2011 as it has a sizeable area under plains. In all districts lying in the plains have a lower female ratio except one hilly district Uttarkashi (Table 6 and Figure 8). Some interesting findings could also be derived from the changing sex ratio as depicted in Figure 6. In all plain districts, there is a gradual increase in the female ratio per 1000 males. The only exception is the district Champawat. However it has either remained moderately the same in hills. The situation is different at the age group of 0-6 which shows lower female children than males in all districts and across all urban areas of the state. Pithoragarh has recorded a minimum of 816 female children per 1000 males in total with a further lowest 724 female children per 1000 males in its urban centers in the age group of 0-6.

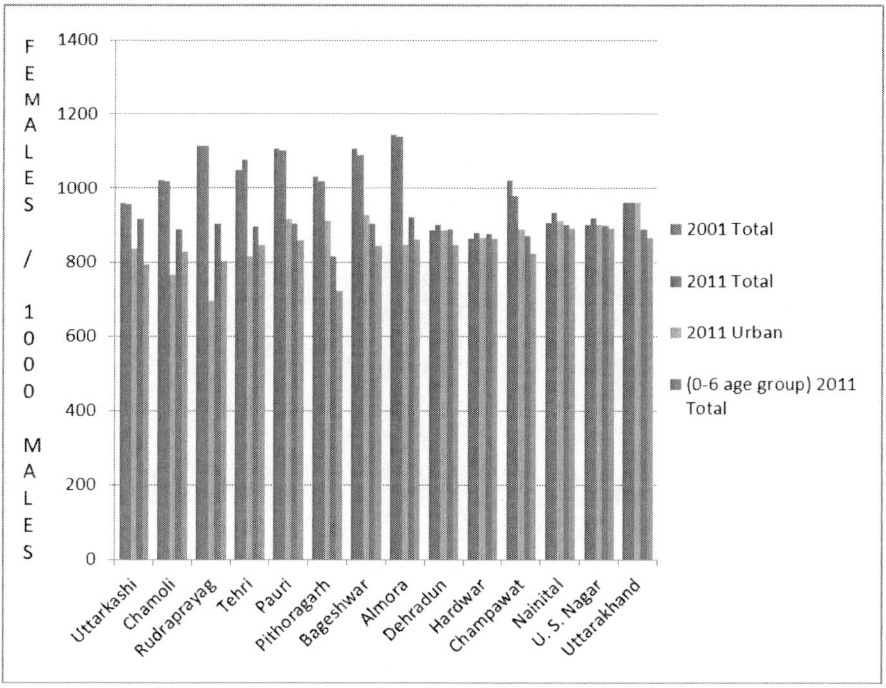

Figure 8. Total, Urban and Child Sex Ratio in Uttarakhand, 2001 and 2011.

Literacy

Beckett (1874) estimated just a 0.5% literacy rate in Kumaun in 1872 which rose to 4.28% in 1881 with a maximum share of males with 96.71% and the least females with 3.29% in literacy rates (Pant and Chand, 2014). Almora and Dehradun were two important centers of education. In the 1901 census, 6.25% literates were enumerated in Uttarakhand which remained almost the same 6.42% in 1911. After independence, literacy rose to 18.73% with male literacy of 32.15% and 4.78% literacy of females. In 1961, it remained stagnant at 18.05% with a slight decrease of 0.68%. However, female literacy increased by to7.33% in 1961. A sudden change was recorded in overall literacy perhaps due to the changing standards of literacy measurement in 1971 with 33.26% and it continued in the following census years.

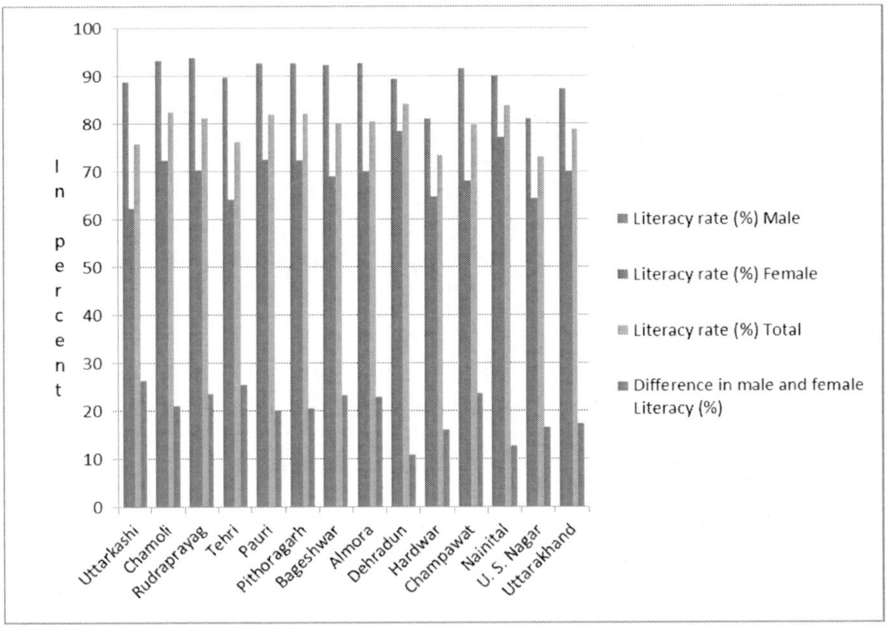

Figure 9. Literacy Rates in Uttarakhand, 2011.

The year 1981 recorded 46.06% maintaining the same trend of increase in 1991 with 57.75% and 72.28% in 2001 and finally 78.82% in the 2011 census. A simultaneous increase was noted both in male and female literacy, the male and female literacy in 2011 were 87.40% and 70.01% respectively.

Table 7. Literacy status in Uttarakhand, 2011

Districts/Region	**Literacy rate (%)**			**Difference in male and female Literacy (%)**
	Male	Female	Total	
Uttarkashi	88.79	62.35	75.81	26.43
Chamoli	93.40	72.32	82.65	21.08
Rudraprayag	93.90	70.35	81.30	23.55
Tehri	89.76	64.28	76.36	25.47
Pauri	92.71	72.60	82.02	20.11
Pithoragarh	92.75	72.29	82.25	20.46
Bageshwar	92.33	69.03	80.01	23.30
Almora	92.86	69.93	80.47	22.94
Dehradun	89.40	78.54	84.25	10.87
Hardwar	81.04	64.79	73.43	16.06
Champawat	91.61	68.05	79.83	23.56
Nainital	90.07	77.29	83.88	12.77
U. S. Nagar	81.01	64.45	73.10	16.64
Uttarakhand	87.40	70.01	78.82	17.40

Source: Census of India, 2011.

A cursory look at the district level distribution of literacy in 2011 revealed that Dehradun ranks first with 84.25% closely followed by Nainital (83.88%). In all, eight districts are falling above 80% of literacy. Of those five districts below 80% are either located in the plain areas or the remotely located Uttarkashi and Tehri districts. Udham Singh Nagar is at the bottom with 73.10% in overall literacy. The ranking in respect to male literacy differs significantly with district Rudraprayag at the top with 93.9% followed by Chamoli (93.4%), Almora (92.86%), Pithoragarh (92.75%), Pauri (9271%), Bageshwar (92.33%), Champawat (91.61%) and Nainital (90.01%). As such, there are also eight districts above 90% of male literacy. District Hardwar (81.01%) and Udham Sing Nagar (81.09%) are at the bottom. In the case of female literacy, Dehradun ranks first with

75.54% and the lowest is in Uttarkashi with 62.35%. There are six districts above 70% of female literacy in Uttarakhand. The increase in female literacy between 2001 and 2011 is noted higher (10%) than male literacy (6%). Another encouraging fact is that the gap between the literacy of two sexes is gradually bridging up from being 23.6% in 2001 to 17.4% in 2011 (Table 7 and Figure 9).

Level of Urbanization

The process of urbanization in Uttarakhand by its intrinsic nature followed the same plain orientation as it was revealed by density and growth variables. There were 86 urban centers in Uttarakhand in 2001 which are now 115 in 2011. The state capital and those located in the foothills show much higher levels of urbanization in comparison to other hilly districts. In hills, the newly emerging urban centers are administrative headquarters and have been consequential upon other institutional infrastructure such as socio-cultural or politico-administrative. There are a majority of lower-class centers in the hills that are characteristically very small in population size and do seldom assume an urban character (Pant et al., 2018, Pant and Chand, 2020). Further in the absence of industrial and large scale commercial activities, these urban centers have evolved due to artificial doze of development given by the state. On the other hand, the causative factors for urbanization in the southern plain parts of the state are favored with more salubrious physic-climatic conditions due to proximity to the plain areas to the further south. This naturally has shown a direct bearing on the socio-economic structure of the people resulting in fairer opportunities of subsistence and survival, commercial transactions and ultimately of development as against the northern districts lying remotely. The situation has been not altered even after the creation of the state and the hill districts are yet waiting for a fairer and balanced approach of development.

Table 8. Urban proportion in Uttarakhand, 2001 and 2011

District /Region	Total Population	Urban Population, 2011		Urban Population, 2001
	Person	Person	%of total	% of total
Uttarkashi	330086	24305	7.4	7.8
Chamoli	391605	59396	15.2	13.7
Rudraprayag	242285	9925	4.1	1.2
Tehri	618931	70139	11.3	9.9
Dehradun	1696694	941941	55.5	12.9
Pauri	687271	112703	16.4	12.9
Pithoragarh	483439	69605	14.4	3.1
Bageshwar	259898	9079	3.5	8.6
Almora	622506	62314	10.0	52.9
Champawat	259648	38343	14.8	30.8
Nainital	954605	371734	38.9	15.0
U. S. Nagar	1648902	586760	35.6	35.3
Hardwar	1890422	693094	36.7	32.2
Uttarakhand	10086292	3049338	30.2	25.7

Source: Census of India, 2011.

The urban population was 25.7% in 2001 which is now 30.2% in 2011. Dehradun district has alone 55.5% share of the total urban population of the state in 2011 which was 52.9% in 2001. It shows that Dehradun has positively influenced by being declared the state capital in 2000 at the time of the creation of the state which was otherwise already a known urban center from the British period. Other districts with higher urban concentration are Nainital (38.9%), Udham Singh Nagar (35.6%) and Hardwar (36.7%). Interestingly, More than 90% of the total urban population is concentrated in these four districts lying in the foothills and plains. There are other districts with an insignificant share of urbanization such as newly formed districts of Rudraprayag (4.1%), Bageshwar (3.5%) and Champawat (14.8%). The situation of old hill districts like Uttarkashi (7.4%), Tehri (11.3%) and Pithoragarh (14.4%) is not much changed in 2011 from that of 2001 (Table 8). Thus, it may be deduced safely that the urban-rural divide is more pronounced in the southern parts by the fact that those urban centers located in the foothills and serving as link towns between hills and plains show the better degree of urbanization in comparison to those located in the hills which continue to be neglected even after the formation of the state.

Human Resource Utilization Pattern

Out of the total population in Uttarakhand, 38.4% of people are registered as workers, while the remaining 61.6% of the population is considered as non-workers (Table 9). The percentage of workers ranges between 30.6% in Hardwar district and 47.9% in Almora district. The hilly districts have a larger working population, while in districts belonging to the plain areas of Uttarakhand; no such distinction could be seen (Figure 10). It is also striking to see that out of the total males, only 49.7% are registered as workers and only 26.7% of the workers are females.

Table 9. Classification of total population/ human resource in Uttarakhand, 2011

District/ Region	Workers (% of total population)			Main Workers (% of total workers)			Non Workers (% of total population)		
	Person	Male	Female	Person	Male	Female	Person	Male	Female
Uttarkashi	47.6	50.0	45.2	81.6	85.0	77.8	52.4	50.0	54.8
Chamoli	46.2	48.4	44.1	63.6	69.9	56.9	53.8	51.6	55.9
Rudraprayag	46.7	45.7	47.5	69.8	72.0	68.0	53.3	54.3	52.5
Tehri	45.3	47.3	43.5	59.2	70.0	48.2	54.7	52.7	56.5
Dehradun	34.3	51.4	15.4	83.8	87.1	71.4	65.7	48.6	84.6
Pauri	39.9	45.1	35.2	60.0	69.2	49.2	60.1	54.9	64.8
Pithoragarh	44.8	47.4	42.2	67.2	70.8	63.2	55.2	52.6	57.8
Bageshwar	47.6	47.2	47.9	63.2	66.1	60.5	52.4	52.8	52.1
Almora	47.9	48.9	47.0	67.4	71.8	63.5	52.1	51.1	53.0
Champawat	38.3	46.1	30.5	63.0	71.6	49.7	61.7	53.9	69.5
Nainital	39.4	52.0	25.9	78.8	84.6	66.3	60.6	48.0	74.1
U. S. Nagar	35.9	51.8	18.6	76.2	83.9	52.9	64.1	48.2	81.4
Hardwar	30.6	49.5	9.1	85.6	88.5	68.1	69.4	50.5	90.9
Uttarakhand	38.4	49.7	26.7	74.1	81.1	60.6	61.6	50.3	73.3

Source: Census of India, 2011.

Table 9 reveals that Pithoragarh district has a minimum of 45.1% male workers, while U. S. Nagar district stands at the top with 51.8%. All in all, nine districts have less than 50% of male workers. The share of female participation in the working category is also shown in Table 9. The percentage of female workers fluctuates from a minimum of 9.1% each in Hardwar districts to a maximum of 47.9% in the Bageshwar district. Table 9 reveals that the share of working women in the region is much lower than that of their male counterparts. It is probably due to the scarcity of jobs in various governmental and non-governmental institutions, shortage of small scale eco- friendly industries and lack of awareness among women to involve her for self-employment. As a result 50.3% of males and 73.3% of females are registered as non-workers.

Out of the total working population (32.4%) in Uttarakhand, only 74.1% of workers are considered as main workers in the 2011 Census. Table 6 shows that Tehri district has a minimum of 59.2% of the main workers and the maximum of 85.6% are found in U. S. Nagar district. About 81.1% of the total male workers are registered as main workers in Uttarakhand, while only 60.6% in this category are females. The share of female workers varies from a minimum of 49.2% in Pauri to a maximum of 77.8% in Uttarkashi district (Table 9). One of the main criteria to classify any settlement belonging to the rural or urban category is occupation, followed by the Census of India. Analyzing the data of different occupational categories in the Uttarakhand, on average, 236.4, 8.6 and 2.7% of the total main workers are registered in the categories of cultivators, agricultural laborers, and workers in household industries, respectively. As per census norms, 52.3% of the main workers are registered as other workers i.e., the tertiary sector (white-collar jobs). The %age of other workers ranges from a minimum of 21.2% in Uttarkashi district to a maximum of 79.8% in the Dehradun district. About 30.8% of the total districts of Uttarakhand have more than 50% of other workers. Similarly, this type of worker also amounts to 61.6% of the total main male workers. Uttarkashi district has the minimum (31.1%) of male other workers, while Dehradun district records the maximum (82.0%). Out of the total female main workers, 28.2% are considered to be in the 'other' category. Rudraprayag district has a minimum of 7.7% of female other workers, while Dehradun district occupies the first rank with 69.8% (Table 10).

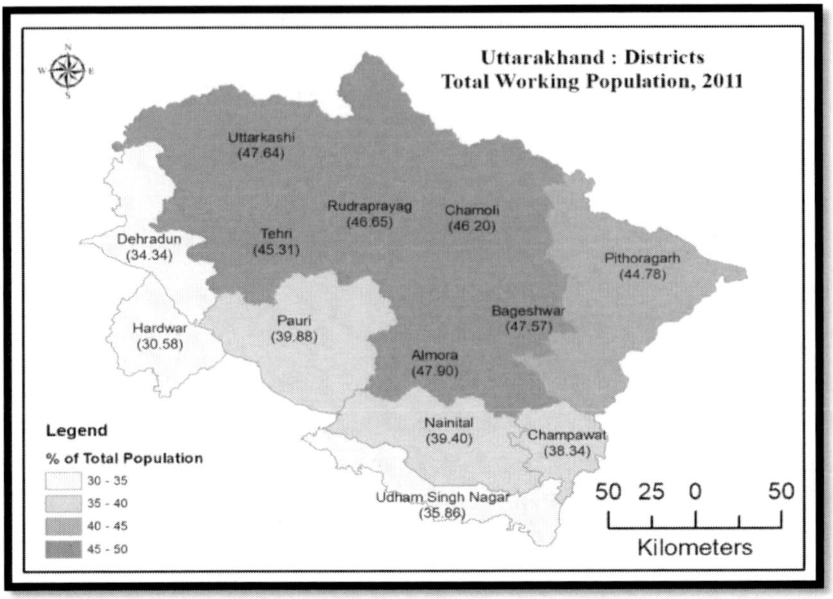

Figure 10. Total Working Human Resource in Uttarakhand, 2011.

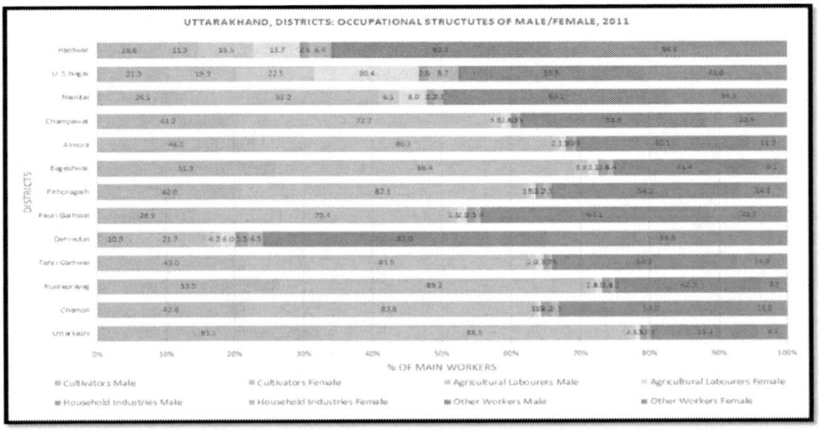

Figure 11. Category wise Occupations in Uttarakhand, 2011.

Table 10. Utilization structure by category of total human resource in Uttarakhand, 2011 (% of main workers)

District /Region	Cultivators			Agricultural Laborers			Workers in Household Industries			Other workers		
	Person	Male	Female	Person	Male	Female	Person	Male	Female	Person	Male	Female
Uttarkashi	75.4	65.1	88.5	1.9	2.1	1.5	1.5	1.7	1.3	21.2	31.1	8.7
Chamoli	60.5	42.8	83.8	0.9	1.0	0.9	2.7	3.2	2.1	35.9	53.0	13.2
Rudraprayag	72.1	53.3	89.2	1.9	1.8	2.0	1.9	2.6	1.1	24.2	42.2	7.7
Tehri	58.8	43.0	81.9	2.2	2.0	2.3	1.3	1.7	0.8	37.7	53.3	14.9
Dehradun	12.4	10.3	21.7	4.2	4.2	4.0	3.7	3.5	4.5	79.8	82.0	69.8
Pauri	45.8	28.9	73.4	2.5	2.5	2.5	2.1	2.5	1.4	49.6	66.1	22.7
Pithoragarh	59.9	42.0	82.1	1.5	1.5	1.5	2.3	2.2	2.3	36.3	54.2	14.1
Bageshwar	69.2	51.9	86.4	3.5	3.9	3.1	2.1	2.8	1.4	25.2	41.4	9.1
Almora	65.7	46.0	86.1	2.0	2.1	1.9	1.4	1.8	0.9	30.9	50.1	11.2
Champawat	51.0	41.2	72.7	3.2	3.3	2.8	1.5	1.7	0.9	44.4	53.8	23.6
Nainital	34.1	26.5	55.2	6.6	6.1	8.0	2.3	2.2	2.5	56.9	65.2	34.3
U. S. Nagar	21.0	21.3	19.3	23.9	22.5	30.4	3.7	2.6	8.7	51.5	53.5	41.6
Hardwar	17.8	18.6	11.3	15.3	15.5	13.7	3.0	2.6	6.4	63.9	63.3	68.6
Uttarakhand	36.4	26.3	62.5	8.6	9.5	6.4	2.7	2.6	2.9	52.3	61.6	28.2

Source*:* Census of India, 2011.

One of the main criteria to classify any settlement belonging to the rural or urban category is occupation, followed by the Census of India. Analyzing the data of different occupational categories in the Uttarakhand, on average, 236.4, 8.6 and 2.7% of the total main workers are registered in the categories of cultivators, agricultural laborers, and workers in household industries, respectively. As per census norms, 52.3% of the main workers are registered as other workers i.e., the tertiary sector (white-collar jobs). The %age of other workers ranges from a minimum of 21.2% in Uttarkashi district to a maximum of 79.8% in the Dehradun district. About 30.8% of the total districts of Uttarakhand have more than 50% of other workers. Similarly, this type of worker also amounts to 61.6% of the total main male workers. Uttarkashi district has the minimum (31.1%) of male other workers, while Dehradun district records the maximum (82.0%). Out of the total female main workers, 28.2% are considered to be in the 'other' category. Rudraprayag district has a minimum of 7.7% of female other workers, while Dehradun district occupies the first rank with 69.8% (Table 10 and Figure 11).

Scheduled Castes Population

The population distribution in the Uttarakhand Region is very uneven due to its undulating topography with the influence of different regional climatic conditions. Mostly population of the Himalaya is concentrated in the plain area of Tarai, Bhabar, Duns, Dwars and river valleys due to their life supportive conditions such as warm climate, abundant water, fertile soil and more suitable area for low-cost infrastructural development while having these adverse conditions, the mountainous part of the region is sparsely inhabited. According to Census 2011, the Scheduled Castes population of Uttarakhand is registered 1892516 people accounting of 18.8% of the total Scheduled Castes population of Uttarakhand (Figure 12). Maximum 27.7% of the total Scheduled Castes population of the Uttarakhand recorded in the Bageshwar district. Pithoragarh district with a population of 120378 persons accounting to 24.9% of the Uttarakhand Scheduled

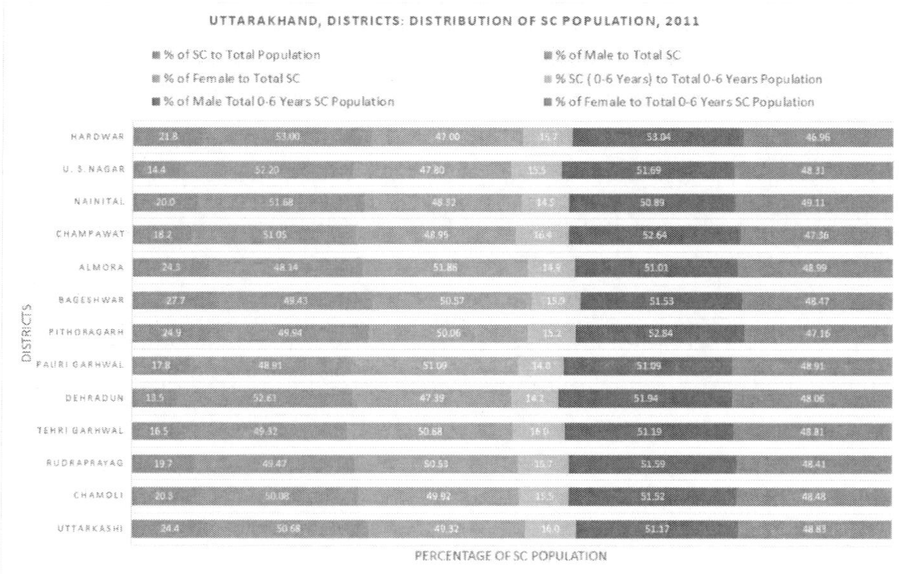

Figure 12. Distribution of Scheduled Castes Population in Uttarakhand, 2011.

Castes population is the second-largest populated districted in the Uttarakhand. Dehradun is the smallest district in the Uttarakhand with a Scheduled Castes population of 13.5% Scheduled Castes population (Table 11).

Scheduled Tribe Population

The population distribution of any geographical unit differs from one place to another. It is mainly governed by the environmental conditions which have both restrictive and permissive condition to human activities. The tribal population of the Uttarakhand mostly concentrated relatively in more inhospitable areas like rugged mountains, extreme climatic zone, dense forest and marshy areas. According to the Census of India 2011 the tribal population of the Uttarakhand is enumerated 2,91,903 persons constitute about 2.89% of the total Uttarakhand population and 10,42,81,034 person are recorded as tribal people in the whole country which is 8.6% of the total population. The Uttarakhand contains 0.28% of the total country's tribal population. The concentration of Uttarakhand tribal population varies 0.14% in Tehri district to 7.46% in U. S. Nagar district. As per Census 2011, the maximum tribal population of the Uttarakhand is found in U. S. Nagar (7.46%) with 2 main tribes (Tharu and Buksa). The second-largest number of tribal people is registered in Dehradun (6.58%) immediately followed by Pithoragarh (4.04%) (Table 11). Remaining districts have less than 4% Uttarakhand tribal population. It is worth to mention that the districts which have more inhospitable geographical areas registered more tribal population. Contrary to this the districts having more level fertile land, irrigational and other infrastructural facilities have less tribal population.

Table 11. Scheduled castes and scheduled tribes population distribution in Uttarakhand, 2011

District/ Region	Scheduled Castes (S.C.) Population			S.C. % of total	Scheduled Tribes (S.T.) Population			S.T. % of total
	Person	Male	Female		Person	Male	Female	
Uttarkashi	80567	40833	39734	24.4	3512	1651	1861	1.06
Chamoli	79317	39718	39599	20.3	12260	6021	6239	3.13
Rudraprayag	47679	23585	24094	19.7	386	217	169	0.16
Tehri	102130	50371	51759	16.5	875	459	416	0.14
Dehradun	228901	120430	108471	13.5	111663	58264	53399	6.58
Pauri	122361	59842	62519	17.8	2215	1174	1041	0.32
Pithoragarh	120378	60111	60267	24.9	19535	9558	9977	4.04
Bageshwar	72061	35623	36438	27.7	1982	971	1011	0.76
Almora	150995	72695	78300	24.3	1281	633	648	0.21
Champawat	47383	24188	23195	18.2	1339	777	562	0.52
Nainital	191206	98824	92382	20.0	7495	3801	3694	0.79
U. S. Nagar	238264	124385	113879	14.4	123037	61758	61279	7.46
Hardwar	411274	217981	193293	21.8	6323	3385	2938	0.33
Uttarakhand	1892516	968586	923930	18.8	291903	148669	143234	2.89

Source: Census of India, 2011.

Table 12. Tribes wise distribution of population in Uttarakhand, 2011

Tribes Name	Household (No)	Population			% of total tribal population
		Person	Male	Female	
Bhotia	10,534	39,106	19,168	19,938	13.40
Buksa	10,587	54,037	27,836	26,201	18.51
Jaunsari	12,645	88,664	46,020	42,644	30.37
Raji	180	690	366	324	0.24
Tharu	15,842	91,342	45,884	45,458	31.29
Total classified	49,788	273,839	139,274	134,565	93.81
Total Unclassified	4,314	18,064	9,395	8,669	6.19
All Schedule Tribes	54,102	291,903	148,669	143,234	100.00

Source: Census of India, 2011.

There are five major scheduled tribes in Uttarakhand. As per the 2011 Census there are 2, 91,903 persons in 54,102 tribal families. Out of the total tribal population maximum of 31.29% are registered as Tharu. Jaunsari (30.37%) stood in the second largest tribe of the state. Buksa tribe has 18.51% of the total tribal population. Only 13.40% of tribal are classified as Bhotia tribe. Out of the total scheduled tribe population 6.19% tribal population has not disclosed their tribe. They are registered as unclassified tribal populations (Table 12).

Migration Scenario

The out-migration of the active workforce from the hilly district of Uttarakhand state is the biggest socio-economic problem these areas face today. Not only the out-migration is increasing year after year, but equally serious is the problem that the people who have once migrated are, most often reluctant to return to their origin; rather they become keen to stay and settle outside. The problem is not only an economic one; it imbibes sociological and psychological ingredients as well. In Uttarakhand, 44.20% of its total population is reported as a migrant in 2011which is higher than the country average (37.47%). Out of the total (4457986) migrants 61.63% rural and remaining 38.37% migrants are urban. About 30% of males are registered as migrants while out of the total female population 58.89% of females are considered migrants (Table 13). Females are more than males because during the marriage a woman moved from their paternal home to her husband's home is considered a factor for migration in the Census.

Out of the total rural population, 44.71% of people are registered as migrants, among the total rural migrants 73.56% are moved from rural to rural and only 26.44% of people moved from urban/cities to rural areas/villages. It is very striking features that females are more than the males among the rural to rural migrants while more males than females are registered as urban to rural migrants. It is due to the male population left their counterpart in the villages/rural areas during his employment duration. After retirement from employment they moved from urban to their native places or nearby cities. Out of the total urbanites 32.67% population is enumerated as urban migrants. Only 27.03% population is moved from rural to urban while 72.97% of people have migrated from urban to urban during the 2011 census. The proportion of male and female migrants in rural to urban and urban to urban is equal. The movement of people with their families is responsible for this equal proportion (Table 13).

Census of India, 2011 has mentioned several reasons for migration, of which six are distinguished separately and remaining is included in other categories (Table 14). Maximum 42.64% of total migrants are migrated due to the marriage reason. Of marriage reason only 1.38% are males and 64.49% females. It is a tradition in Uttarakhand where females have to move from parental home to husband's home after completion of the marriage ceremony. In our opinion marriage reason may not be considered because it is a social compulsion in which females are exchanged between the males of different villages or cities. Three, employment, business and education are the main factors for migration in Uttarakhand.

Table 13. Migration pattern in Uttarakhand, 2011

Region	Total/Rural/Urban	Last residence Rural/Urban	Migrants					
			P	%	M	%	F	%
Uttarakhand	Total	Total	4,457,986	44.20*	1,543,627	30.04	2,914,359	58.89
	Rural	Total	2,747,243	61.63$	732,742	47.47	2,014,501	69.12
	Urban	Total	1,710,743	38.37$	810,885	52.53	899,858	30.88
	Total	Rural	3,146,237	44.71**	945,682	26.87	2,200,555	62.55
	Rural	Rural	2,314,424	73.56#	542,993	57.42	1,771,431	80.50
	Urban	Rural	831,813	26.44#	402,689	42.58	429,124	19.50
	Total	Urban	996,128	32.67***	453,833	28.04	542,295	37.91
	Rural	Urban	269,261	27.03##	120,916	26.64	148,345	27.36
	Urban	Urban	726,867	72.97##	332,917	73.36	393,950	72.64

Source: Census of India Provisional, D-5 Migrants by Place of Last Residence.

Note:

* % of total population

** % of total rural population

*** % of total urban population $% of total migrants# % of total rural migrants ## % of total urban migrants and %age for Male and Female has been calculated in similar method.

Table 14. Reasons for migration in Uttarakhand, 2011

Total/Rural/ Urban	**Last residence Rural/Urban**	**Reasons for migration (% of total migrants)**								
		Work/Employment			Business			Education		
		P	M	F	P	M	F	P	M	F
1	2	6	7	8	9	10	11	12	13	14
Total	Total	14.90	38.31	2.49	0.52	1.14	0.18	3.11	5.49	1.85
Rural	Total	10.87	35.25	2.00	0.30	0.84	0.11	2.16	4.64	1.26
Urban	Total	21.36	41.08	3.59	0.86	1.42	0.35	4.64	6.26	3.18
Total	Rural	14.52	43.26	2.17	0.37	0.97	0.11	2.79	5.62	1.57
Total/Rural/ Urban	Last residence Rural/Urban	Reasons for migration (% of total migrants)								
		Work/Employment			Business			Education		
		P	M	F	P	M	F	P	M	F
1	2	6	7	8	9	10	11	12	13	14
Rural	Rural	10.35	38.29	1.78	0.26	0.80	0.09	1.94	4.73	1.08
Urban	Rural	26.14	49.98	3.78	0.69	1.19	0.21	5.15	6.83	3.57
Total	Urban	18.41	35.88	3.79	0.95	1.61	0.40	4.44	6.02	3.11
Rural	Urban	17.57	33.97	4.20	0.65	1.14	0.25	4.32	5.44	3.41
Urban	Urban	18.72	36.57	3.63	1.07	1.78	0.46	4.48	6.23	3.00

Total/Rural/ Urban	**Last residence Rural/Urban**	**Reasons for migration (% of total migrants)**											
		Marriage			Moved after birth			Moved with household			Others		
		P	M	F	P	M	F	P	M	F	P	M	F
1	2	15	16	17	18	19	20	21	22	23	24	25	26
Total	Total	42.64	1.38	64.49	3.62	6.25	2.22	26.30	32.88	22.81	8.92	14.54	5.94
Rural	Total	55.65	2.20	75.09	2.33	5.22	1.28	20.41	34.39	15.32	8.27	17.46	4.93
Urban	Total	21.74	0.63	40.76	5.68	7.19	4.33	35.76	31.51	39.59	9.96	11.91	8.21
Total	Rural	51.00	1.74	72.17	1.79	3.60	1.01	23.80	33.95	19.43	5.73	10.86	3.53
Rural	Rural	61.60	2.60	79.69	1.63	4.19	0.85	18.36	35.15	13.21	5.86	14.25	3.30
Urban	Rural	21.51	0.59	41.13	2.22	2.79	1.68	38.93	32.32	45.13	5.37	6.29	4.50
Total	Urban	23.99	0.72	43.47	6.40	8.35	4.77	35.68	33.67	37.36	10.13	13.75	7.10
Rural	Urban	24.23	0.78	43.35	3.75	4.72	2.96	39.14	38.08	40.00	10.34	15.87	5.83
Urban	Urban	23.91	0.70	43.52	7.38	9.67	5.45	34.39	32.07	36.36	10.05	12.98	7.58

Source: Census of India Provisional, D-5 Migrants by Place of Last Residence.

Out of the total migrants, 14.9% population has moved for getting work or employment from their residence. Out of which 38.31% are males and only 2.49% of females are included in this category. Less number of females is in this category because professional and technical literacy is fewer than males with not promoting females for employment by society. Only 10.87% of rural migrants are migrated for employment while 21.36% are urban who moved for getting work from their residence.

The proportion of females is less than their counterparts. Table 14 shows that rural to rural (10.35%) migrants are lower than urban to rural (26.14%) in search of employment. It is due to the rural areas have more possibilities for expansion of new establishments such as industries, institutions, developmental works, etc. As a result more people moved from less spacious urban areas to more spacious rural areas.

Migration for business is not a more effective factor in Uttarakhand where the meager proportion of migrants is moved from their original residence. Due to the absence of an industrial and viable agricultural environment in the Uttarakhand, education is the only medium to get employment in or outside the state. Unfavorable conditions for employment generation and favorable conditions for Europeans, they established more educational institutions for their children. As a result, Almora, Pauri, Pithoragarh, Mussorie and Dehradun were the hub of educational centers and well off children get a good education and in the present time, maximum village and cities have lower to higher educational institutions. Therefore there is no need to migrate for education in the present decade. Only 3.11% of people have migrated for education in 2011. Among them only 5.49% of males and 1.85% are females. Urban migrants are more than rural migrants who moved for education. About 26.3% of people are moved with households in the region. Out of the total rural and urban migrants 20.41% and 35.76% migrants moved with household respectively. Approximately one-third of migrants are in urban to rural (38.93%), rural to urban (39.14%) and urban to urban (34.39%) moved from their earlier residence to new destinations with household. Table 14 portrays the reasons for migration in Uttarakhand in detail. District-wise data for migration is not available yet.

CONCLUSION

The study area extends from below 300m elevation in south to more than 7000m in north; as a result the Uttarakhand has several altitudinal zones that produce varied climatic conditions which also determine the forest and soil types of the region. These natural important factors are responsible for uneven distribution and growth of the population in the Uttarakhand. The study reflects that 90% of hill area has only 48% population and 10% plain area has 52% state's total population. More than two-thirds of urban people are residing in almost plain region- Tarai, Bhabar and Dun areas of Uttarakhand where more than 70% of the state's urban centers are located. Factors like rugged terrain, harsh climatic conditions, stony less fertile soil, etc. are responsible for scattered distribution and lower growth of population in the mountainous part of Uttarakhand.

Uttarakhand region presents very distinct demographic characteristics determined by the local relief and the harsh climate together with a host of other physical as well as cultural factors. There are areas with no population in the Trans Himalaya to densely populated areas along the foothills. The distribution pattern of the population not only differs from the rest of the country but greatly varies within the states and districts belonging to one part to another part of the region. As one proceeds from the southern Tarai–Bhabar tract to Siwalik, Lesser Himalaya; and Greater Himalaya and Trans Himalayan region of the state, the population decreases markedly by the topography and environmental conditions.

To sum up, the Uttarakhand state has witnessed a 405% growth in its total population from 1901 to 2011. But this increase has not been spatially uniform. The concentration of population has grown only in the areas already populated specially in the districts lying along the southern margins of the state. The decadal growth has been quite upcoming apparently, slowly coming up to 1951 making an abrupt uplift in 1981. The emergence of the state in 2000 has further confirmed the foothill region to ascertain such a high growth trend. In 2011, about 52% population lives in Hardwar, Dehradun, and Udham Singh Nagar alone and the rest 48%

population lives in ten districts. These three districts also share the maximum population in the whole Indian Himalayan region. Hardwar is at the top (4.1%), followed by Dehradun (3.62%) and Udham Singh Nagar (3.51%). A comparison of 2001 and 2011 data between plain and hill districts of the Uttarakhand state revealed that the population of plain districts has grown from 4.7 million to 6.2 million with a net increase of 31%. On the other hand the hill districts have grown from 3.7 million to 3.9 million with a net increase of 3%. District Almora (-1.64) and Pauri (-1.41) have even shown negative growth. The districts lying in the hills have the lowest density and do show the more insignificant or even nil change in concentration of population which is consequential upon the existent physical-climatic environment therein. Although the overall literacy condition in the hills is better, it is here that the gap between the literacy of the two sexes is higher which was 36.9% in 2001 and 26.43% in 2011. Another encouraging fact is that the gap between the literacy of two sexes is gradually bridging up from being 23.6% in 2001 to 17.4% in 2011. The process of urbanization in Uttarakhand by its intrinsic nature followed the same plain orientation as it was revealed by density and growth variables. Dehradun district has alone 55.5% share of the total urban population of the state in 2011 which was 52.9% in 2001.

The problem of migration is not only an economic one; it imbibes sociological and psychological ingredients as well. In Uttarakhand, 44.2% of its total population is reported as migrants in 2011. Out of the total migrants, 61.63% rural and remaining 38.37% of migrants are urban. About 30% of males are registered as migrants while out of the total female population 58.89% of females are considered migrants. Study shows that females are more than the males among the rural to rural migrants while more males than females are registered as urban to rural migrants. It is due to the male population left their counterpart in the villages/rural areas during his employment duration. After retirement from employment they moved from urban to their native rural places or outskirts of cities. The analysis of out-migration in Uttarakhand shows that the pull factor is dominant in district situated in plain areas like Nainital, U. S. Nagar, Hardwar and Dehradun being infra-structurally and socio-economically more developed than their surrounding hilly districts. If the same trend of our migration continues, it is quite possible that in the coming years, the very base of the hill economy might collapse. The mainstay of the economy i.e., agriculture and allied activities is likely to come to standstill.

Hills are already on the verge of population collapse. The loss of population if viewed at the sub-district level presents a more alarming situation. In the hills, out of 78 sub-districts, 16 (20.06%) sub-districts have registered a decline in the population from 2001 to 2011. If the same trend continues, 29 (37.2%) sub-districts would further decline their population in 2021. On the positive side, there are 18 (23%) sub-districts with more than 20.01% growth, but most of them are in the plain area. However these imbalances would further cause a socio-political rift between the people residing in plains and hills of Uttarakhand.

REFERENCES

Anonymous, 2001. *Census of India, Primary Census Abstract*, 2001. (http://www.censusindia.gov.in).

Anonymous, 2011. *Census of India, Primary Census Abstract*, PCA0500 to 0513_2011_MDDS Uttarakhand. (http://www.censusindia.gov.in).

Anonymous, 2011. *Census of India, 2011* pca_state_distt_st 2011(http://www.censusindia.gov.in).

Anonymous, 2011. *Census of India, 2011*, Census of India, pca_state _distt, 2011. (http://www.censusindia.gov.in).

Anonymous, 2011. *Census of India, 2011*, Census of India, pca_state _distt_sc. 2011 (http://www.censusindia.gov.in).

Anonymous, 2011: *Census of India, 2011* Provisional, D-5 Migrants by Place of the Last Residence. (http://www.censusindia.gov.in).

Anonymous, 2015. : *Atlas Forest Types of India*, 2011 Quoted by ISFR, 2015: 253-254p.

Anonymous, 2017. *India States of Forest Report* (ISFR), 2017 page- 309 - 311.

Chand, R. (2013). Labour Migration as a Livelihood Strategy in Far East Bhutan: A Case Study of a Marginal Bhutanese Community, *Hrvatski Geografski Glasnik*, 75(2): 41–57.

Chand, R. (2017). Social Ecology of Immigrant Population and Changing Urban Landscape of Thimphu, Bhutan, *Journal of Urban and Regional Studies on Contemporary India* 4(1): 1–12.

Chand, R., Taragi, R. C. S., and Pant, B. R. (2016). Changing the Demographical Profile and Migration Pattern in Uttarakhand. In *Uttarakhand: Nature, Culture, Biodiversity* edited by Negi, Chandra Singh, Winsar Publishing Co., Dehradun: 390-403.

Ghildyal, B. P. (1951). Soils of the Garhwal and Kumaun Himalaya. In *The Himalaya: Aspects of Change*, edited by Lall, J. S. and Moddie, A. D., Oxford University Press New Delhi: 120-137.

Joshi, S. C., Joshi, D. R. and Dani D. D. (1983). *Kumaun Himalaya: A Geographic Perspective on Resource Development*. Gyanodaya Prakashan, Nainital (U. P.) India. 324p.

Pant, B. R. (2013). Growth and Distribution of Population in the Indian Himalaya. *The Geographer*, 60(2):76-89.

Pant, B. R. (2012). Demography of Indian Himalayan Region in Census, 2011, *ENVIS Bulletin on Himalayan Ecology*, 20: 1-20.

Pant, B. R. (1994). *Food and Nutrition: A Study of Himalayan Region*, Anmol Pub. Pvt. Ltd, New Delhi: 195p.

Pant, B. R. (1998). *Geo-Economic Environment and Nutrition*, Anmol Pub.Pvt Ltd New Delhi: 173 p.

Pant, B. R. (1988). *Land Utilization in Kotadun, Kumaun Himalaya.* Unpublished PhD Thesis Submitted to the Kumaun University, Nainital: 288p.

Pant, B. R. (1995). The Central Himalaya: Geographical Outlook. In *Glimpses of Central Himalaya: A Socio-Economic & Ecological Perspective*, Edited by Pant B. R. and Pant M. C. Radha Publications, New Delhi: 3-86.

Pant, B. R. and Chand, R. (2020). Demographic Characteristics of Urban Centres and Urbanization in Uttarakhand. *Journal of Urban and*

Regional Studies on Contemporary India. Vol. 6 (2): 1-20 HINDAS National Institute for the Humanities, Hiroshima, Japan.

Pant, B. R. and Chand, R. (2014) *Uttarakhand: Jansankhyatmak Jhalak*, (in Hindi) (*Demographic Glimpses of Uttarakhand*) PAHAR Published a Booklet on the occasion of Fifth Askot –Arakot Abhiyan, Nainital: 2-23.

Pant, B. R. and Chand, R. (2013). 2011 Mai Bhartiya Himalaya (in Hindi) (Indian Himalaya in Census, 2011), *Journal of PAHAR*, Nainital-18:22-54.

Pant, B. R., Chand, R. and Taragi, R. C. S. (2018). Urbanization in the Indian Himalayan Region. In *Environment, Resources and Development of the Indian Himalaya*, edited by Rawat, M. S. S. Singh Vir, Upreti Paritosh, Agarwal Apeksha and Singh Shweta. Transmedia Publications Srinagar (Garhwal): 79-110.

Pant, B. R., Chand, R. and Mehta B. S. *Uttarakhand: Jansankhya Paridrashya Evam Parivartan* (in Hindi) (*Uttarakhand: Population Scenario and Changes*), PAHAR POTH, Nainital in press.

Valdiya, K. S. (1988).Geology of Kumaun: An Outline. In *Kumaun Land and People, edited by* Valdiya K. S., Gyanodaya Prakashan, Nainital: 37-72.

Valdiya, K. S. (1980a). *Geology of Kumaun Lesser Himalaya*, WIHG, Dehradun: 301p.

Valdiya, K. S. (1980b). Stratigraphic Scheme of the Sedimentary Units of the Kumaun Lesser Himalaya, In *Stratigraphy and Correlation of Lesser Himalaya Formations,* edited by Valdiya, K. S. and Bhatia, S. B.. HPC, New Delhi & WIHG, Dehradun: 7-48.

BIOGRAPHICAL SKETCH

B. R. Pant

Affiliation: Professor of Geography, M. B. Govt. P.G. College Haldwani, (Nainital)

State/Country: Uttarakhand, India

Education: MA, PhD Geography

Research and Professional Experience: Last 35 years

Professional Appointments: I-As Anskalik/ part time lecturer, Assistant professor, Associate Professor and Professor.

II- Worked as Registrar and Examination Controller in Uttarakhand Open University, Haldwani (State Open University)

III- Life Member of various professional / academic societies such as Geographical Society of India, National Association of Geographers India, People's Association of Himalayan Area Research, Central Himalayan Environmental Association, Punjab Geographical Society, etc.

Honors: Bharat Jyoti Award for Meritorious Services Outstanding Performance and Remarkable Role by India International Friendship Society, New Delhi

Publications from the Last 3 Years:

Research Papers in National & International Journals

1) Kumar S. and B. R. Pant, 2019. Process and Pattern of Urban Growth: A Case Study of Muzaffarnagar City. *Journal of Emerging Technologies and Innovative Research* (JETIR) Vol. 6, Issue 6: 69-77 (ISSN-2349-5162) www.jetir.org.
2) Pant B. R. 2016. Demographic Profile and Nutrition Status of Women in Uttarakhand. *ENVIS Bulletin: Himalayan Ecology*, GBPIHED, Kosi- Katarmal, Almora, Vol. 24, 96-103 pp .ISSN:0971-7447(Print) 2455-6815(On line).
3) Singh, T. B. and B. R. Pant. 2016. Demographic Structure of Haldwani Agglomeration, Distt. Nainital, Uttarakhand.

Vidyawarta, International Multilingual Research Journal, 16 (5): 62-68. ISSN-21199318 (Impact Factor 4.014).

4) Pant B. R. 2016. Pattern of Population Growth During 1901 to 2011 in the Indian Himalayan Region. *International Journal of Research in Commerce, Economics and Management (IJRCM),* 6 (4): 13-27 (ISSN 2231-4245).
5) Pant B. R. 2016. Literacy Pattern Among Scheduled Castes Population in the Indian Himalayan Population, *Printing Area, International Multilingual Research Journal*, Vol. 01 (13): 41-54 (Impact Factor, 3.024).
6) Kumar Sushil and B. R. Pant, 2020. Assessment of Quality of Life and Economic Status in the Slums of Mujaffarnagar. *International Journal of Current Research* Vol.12 (2): February, 2020:10105-10108ISSN-0975-833X. doi: https://doi.org/10.24941/ijcr.37959.02.2020.
7) Pant, B. R. and R. Chand 2020. Characteristics of Urban Centres and Urbanization in Uttarakhand. *Journal of Urban and Regional Studies on Contemporary India*. Vol. 6 (2) ISSN- 2188-5222 (HINDAS National Institute for the Humanities, Hiroshima, Japan: 1-20.

Research Papers in Books/Proceedings

1) Pant, B. R. and R. Chand 2019. Demographic Characteristics of Urban Centres and Urbanization in Uttarakhand. In: K. C.Purohit, Anita Rudola and Pranjali Purohit (eds.) *The Himalayas: At the Cross Road of Environment and Development*. Winsar Publishing, Dehradun: 101-135 ISBN: 978-93-83830-52-8.
2) Pant B. R. and R. Chand and R. C. S. Taragi. 2018. Urbanization in the Indian Himalayan Region, In M. S. S. Rawat, Vir Singh, Paritosh Upreti, Apeksha Agarwal and Shweta Singh (eds.) *Environment, Resources and Development of the Indian Himalaya.* Transmedia Publication Srinagar (Garhwal):79-110. ISBN: 978-81-936527-1-8.

3) Pant B. R. and R. Chand, 2018. Tribal and Non-Tribal Population in the Indian Himalayan Region: Some Aspects. In K. D. Sharma and K. Surjit Singh (eds.) *Massive Urbanisation: Town Planning, Pedagogy & Research,* ISPER (Institute for Spatial Planning and Environment Research), Panchkula (Haryana): 161-182.ISBN 978-81-927014-0-4.
4) Pant B. R. and S. Pal, 2017. Socio - Cultural and Nutritional Environment of a Marginal Community of Uttarakhand, India – A Case Study of the Tharu Tribe. In R. Chand, Etienne Nel and Stanko Pelc (eds.) *Perspectives on Geographical Marginality, Vol. II, Societies, Social Inequalities and Marginalization: Marginal Regions in the 21st Century*. Springer, International Publishing AG Switzerland. ISBN - doi 10.1007/978-3-319-50998-3_14: 213-231.
5) Pant B. R. and Ghananand Palariya. 2017. Uttarakhand ka Janjateey Janankikeya Adhayan. In Harit Kumar Meena (ed.) *Bhartiya Aitihasik - Sanskritik Parampara aur Adivasi Vimars* S. K. Book Agency, New Delhi:20-49. ISBN: 978-93-8315-896-6.
6) Pant B.R. and T. B. Singh, 2018. Age Wise Demographic Scenario of Uttarakhand and India: With Special Reference to Distribution, Sex Ratio and Literacy. In H. S. Bisht (ed.) *Ageing and Healthcare in India: Issues and Challenges*. Indu Book Services, New Delhi: 17-41. ISBN: 978-93-86754-29-5.
7) Pant B.R. and R. Chand (in press). The Crises of Ethnicity and Urbanism in India (Chapter 14). In: Hiromi Tanaka, Chompoonuh K. Perpoonwiwet and Divya U. Joshi (eds.) *Gendered Cityscapes: Gender Identity, equity and marginalization in Urban Asia*. Rawat Publishers Jaipur (in press).

In: India
Editors: J. Lessard and O. Rivière
ISBN: 978-1-53618-413-6

Chapter 2

The Recovery of Biodiversity in Commercially Logged Rain Forests in Sri Lanka

A. H. Magdon Jayasuriya[1,*], PhD
and S. D. Abayawardana[2]
[1]EML Consultants (Pvt) Ltd, Kotte, Sri Lanka
[2]Integrated Development Association (IDEA)

Abstract

As nature, the forest too is in a state of flux and it possesses innate ability to thwart impacts, both natural and anthropogenic. The majority of the Tropical Rain Forests in Sri Lanka have suffered selective logging, more intensely five to six decades ago. The Sinharaja and Kanneliya are two such largest remaining lowland rain forest areas in Sri Lanka. The National Conservation Review (NCR) that was conducted between 1991 and 1996 to evaluate all remaining natural forests used Gradient-directed transect (gradsect) method to sample these forest blocks to assess species diversity. The Biodiversity Monitoring Survey (BMS) was conducted in

* Corresponding Author's E-mail: magdonj@emlconsultants.com.

2008 to determine the changes in the status of plant biodiversity in Sinharaja and Kanneliya forests since the NCR survey was conducted in these sites in 1992, i.e., over sixteen years prior to the BMS. It was evident that Kanneliya, along with two other contiguous forest areas, possess richest plant diversity with highest number of species and endemic species in the island. Positive trends in some important biodiversity characteristics after sixteen years were noted. Changes in succession status in both sites, i.e., increase in primary species and decrease in combined pioneer and secondary species clearly indicated that forests are successfully regenerating accompanied by an improvement in biodiversity quality. In Sinharaja the recovery in sites that were abandoned after slash and burn (chena) cultivation was clearly evident by similar observations. In Kanneliya, after nearly two decades of recess since timber exploitation, observations indicated recovery of forest stature accompanied by the appearance of emergents and reproductively mature canopy trees gradually bringing back the original seed bank into the forest dynamics. The enrichment planting with Mahogany in selectively logged plots in Kanneliya during early 1990's was a proven failure. The above findings indicated that the Tropical Lowland Rain Forests are capable of recovery, in terms of biodiversity and stature, if sufficient recess is provided following land use practices such as selective logging and chena cultivation in spatial and temporal contexts.

Keywords: biodiversity monitoring, forest recovery, forest succession, selective logging, tropical rain forest

1. Introduction

1.1. Sinharaja Biosphere Reserve and Natural World Heritage Site

Sinharaja Biosphere Reserve (hereafter Sinharaja) is located in the southwest of Sri Lanka at 6^0 21-26' N and 80^0 21-34' E in the southern part of the wet zone in Sri Lanka. It lies between the tributaries of the Kalu Ganga river (in the North) and the Gin Ganga river (in the South) and extends within the administrative districts of Galle and Matara (Southern Province) and Ratnapura (Sabaragamuwa Province). At present, Sinharaja is a large block of lowland rain forest in Sri Lanka and it is administered by the Forest Department [1, 2, 3]. Sinharaja spreads over an elevation range of 200 to 1,150 m above mean sea level and the terrain is generally rolling, with a series of ridges and valleys, while some parts of the Eastern section of the forest consist of plains. There are about nine prominent peaks ranging from 575 m to 1,170 m of which the Northeastern peak of Hinipitigala is the highest. The mean annual rainfall varies between 3,600 and 5,000 mm. There is no distinct dry season and the mean annual temperature ranges between 19^0 and 27^0C. The Sinharaja forest is one of the most important watersheds in the island, with several streams draining into the 'Kalu Ganga' and 'Gin Ganga' rivers (Figure 1).

Prior to 1972, Sinharaja was protected largely by its inaccessibility. Between 1971 and 1977 the western part of the Sinharaja became a production forest, selectively logged for plywood. Concurrently, 40 ha at Warukandeniya in the south central part and 1800 ha in the eastern part were designated as National Man and Biosphere (MAB) Reserves. Logging activity ceased in 1978 due to intense protests from nature conservationists and the whole extent of 8,500 ha, comprising 65 percent high forest, and 34 percent fernlands and secondary forest, was declared an International Biosphere Reserve [4]. In the 1980's, an additional 2,687 ha of lower montane forests on the eastern end at Thangamale and Morningside were included in the Sinharaja reserve making a total of 11,187 ha. The Forest Department established a live boundary for MAB Reserve by planting *Pinus caribaea* in 1985. The Sinharaja National Heritage Wilderness Area (NHWA) was declared under the NHWA Act, and Gazetted on the 21st of October 1988. It refers to NHWA Sinharaja Map of 20th October 1988 prepared by the Surveyor General as a NHWA. In 1990, UNESCO recognized it as a Natural World Heritage site [5, 6].

The 11,187 ha of legally protected forest is the "core area" of the Biosphere Reserve, while there is an external buffer zone outside this area [7]. The management plan for Sinharaja recommends strategies to ensure maximum protection of the legally protected core area, while livelihood and traditional practices are enabled in the buffer zone as long as it does not damage the protected forest. Although research indicates that a 3 km buffer zone was set up to facilitate economic activities and to prevent encroachment [8], there is no legally defined buffer zone specified for the NHWA [8].

1.1.1. Research

Early research on the flora of Sinharaja was limited to the gathering of herbarium specimens even as far back as 1855 [9, 10, 11]. When logging was stopped in 1977, biological research was initiated to examine biodiversity and its distribution to support effective conservation management of the forest. The enhancement of the accessibility to the reserve and network of logging trails and skid roads and the improvement of other logistics attracted researchers to conduct research at Sinharaja. Among the faunal studies, the ecology of mosquitoes, small mammals, bird flocks and frogs have received much attention. The next phase emphasized targeted research on the regeneration of timber and non-timber species in selectively logged sites, reproductive biology of economically important plants harvested directly from the wild, genetic diversity of Dipterocarpaceae, eco-physiology and silviculture of selected canopy dominant species, restoration and rehabilitation ecology, natural resource economics, ethno-botany and rural sociology of the surrounding villages. In the early 1990s, long-term forest dynamic studies in the primary forests and mixed species trials in a buffer zone *Pinus caribaea* plantation were initiated [2]. Concurrently, the plant taxonomists engaged in the Flora of Ceylon Project in 1970s and 1980s have gathered many specimens that were cited in the Revised Handbook to the Flora of Ceylon [9]. These specimens have been deposited at the National Herbarium (NH). Subsequently, Jayasuriya & Balasubramaniam, supported by the National Science Foundation (NSF) conducted a survey of the flora and deposited over one thousand herbarium vouchers at the NH [10].

However, most of these studies have been conducted in the western sector of the reserve, approached via Kudawe, while the eastern, southern and northern parts have received relatively less attention. Subsequently, the National Conservation Review (NCR), considered as a comprehensive biodiversity survey at national scale, was conducted during 1991-1996 [11]. The western, eastern, northern and southern sectors of the Sinharaja were comprehensively examined during this survey and 337 species of woody flowering plants in 71 families and 191 genera were recorded. Of these, 192 species were endemic (57%), while 10 species were apparently recorded only from Sinharaja. The faunal species comprised of 13 mammals, 13 reptiles, 44 birds, 11 amphibians, 15 mollusks and 11 butterflies. Almost 50% of the faunal species were endemic. The data gathered and the botanical specimens collected during above surveys in the Sinharaja made notable contributions to the knowledge of the flora and for updating the National Red List of Sri Lanka [9, 12].

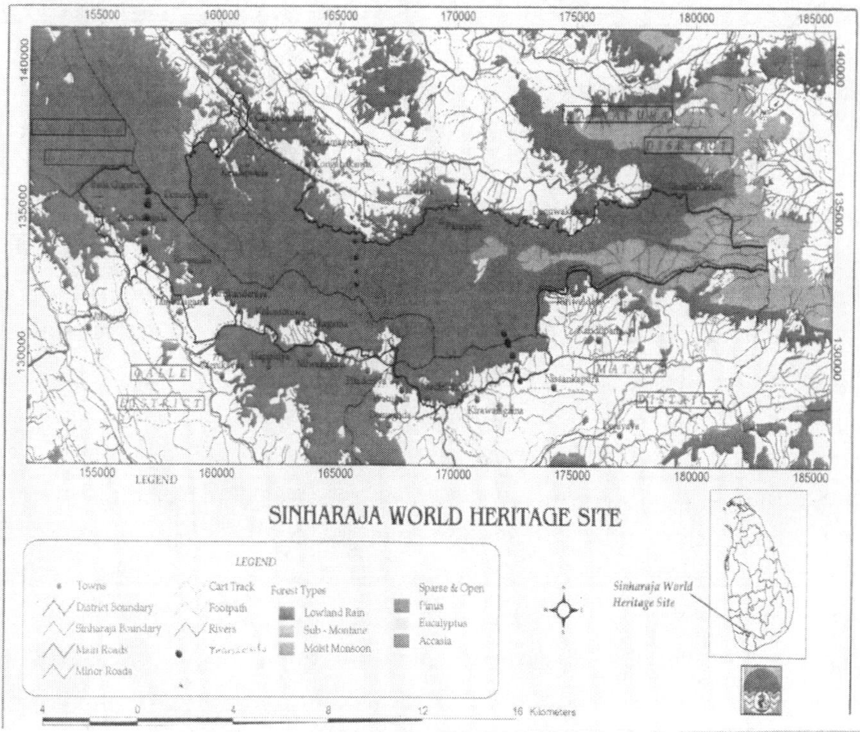

Source: Forest Department.

Figure 1. Map of Sinharaja World Heritage Site – some NCR gradsects/plots are shown as black dots.

1.2. Kanneliya-Dediyagala-Nakiyadeniya Biosphere Reserve

Kanneliya (6114.4 ha in Galle District), Dediyagala (3789.9 ha in Matara District) and Nakiyadeniya (2292.1 ha in Galle District) are contiguous Forest Reserves forming a single large forest unit (KDN Complex) that comprises of a total area of 12,196.4 ha situated in the southwest of Sri Lanka at 6^0 09-18'N and 80^0 19-27'E in the southern part of the wet zone (Figure 2). The average annual rainfall is about 4,445 mm received mainly during the Southwest monsoon season (from mid-May to end September) and the mean monthly temperature is around 27^0C [13]. The KDN Complex is ranked as the most important natural forest in the Southern Province for the protection of headwaters of river systems, as it has the highest number of streams flowing through it (111 streams). The Kanneli Ela and Nannikita Ela catchments have a high proportion of steep slopes and high levels of forest cover indicating their superiority as catchments [3]. The mean elevation of the KDN complex is about 180m varying between 60 to 425m. Kanneliya consists of several parallel ridges and valleys aligned northwest – southeast, while the ridges in Nakiyadeniya and Dediyagala are lower and often broken into isolated, rounded hills, all rising to about 230 m [14].

KDN has been subjected to timber harvesting from pre-independence period up to over four decades ago. Since 1950, controlled exploitation by coupes managed under the selection system was undertaken. However, administrative and other difficulties had led to the 'high grading' and illegal felling within the reserves [14, 15]. Therefore, even before 1961 there were large areas of advanced secondary forests within the KDN Complex. In 1968, the Plywood Corporation, on a lease obtained from the Forest Department, undertook a mechanized logging project for intensive timber harvesting. With the introduction of a moratorium on all logging within the wet zone in 1990, KDN was finally saved from timber exploitation [13]. However, this disastrous activity left behind a network of degraded hill slopes, colonizing gaps and some un-logged patches on steeper slopes. Furthermore, other factors such as encroachment for plantation crops, e.g., tea, have caused conversion of forest tracts into secondary forests and degraded lands that are in different stages of regeneration.

The Kanneliya forest was selected to be conserved under the World Bank funded *Conservation and Sustainable Use of Medicinal Plants Project* and is considered to be a habitat for important medicinal plant species.

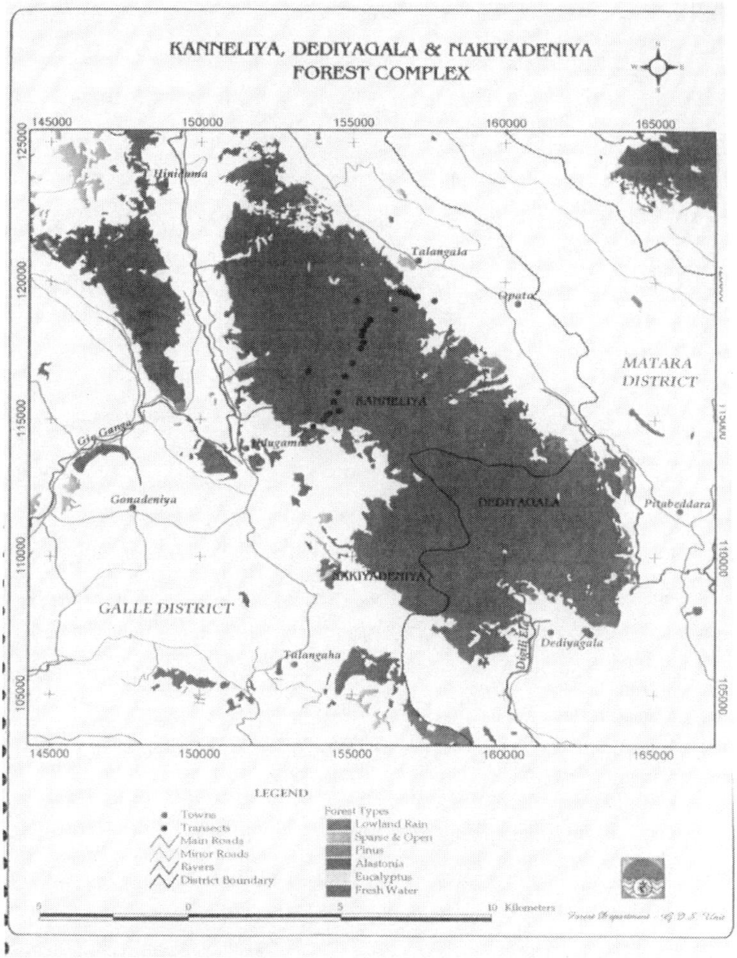

Source: Forest Department.

Figure 2. Map of Kanneliya – Dediyagala – Nakiyadeniya Complex – some NCR gradsects/plots are shown as black dots.

The KDN Complex is located in Floristic Region 7 (Southern Sinharaja-Hiniduma-Kanneliya), which comprises the outer range of the southern block of hills (16). This is considered to be an area of exceptional endemism, as some 17% of the wet zone lowland endemic flora is confined to the south-western front-range forests. The NCR survey, conducted between 1991 and 1996 [11], revealed that the KDN Complex is one of the eight contiguous forests that accounts for a higher proportion of the woody plants and animal diversity in the island. This region is identified as the floristically richest area in South Asia. Considerable information on flora of the KDN Complex has been generated by studies conducted by Peeris [14] and Sinhakumara [15]. Concurrently, the plant taxonomists engaged in the Flora of Ceylon Project in 1970s and 1980s have gathered many specimens that were cited in the Revised Handbook to the Flora of Ceylon [9, 10, 11]. The collections made during NCR survey also have made an important contribution to the knowledge on the flora of the KDN Complex.

1.3. GEF Medium Grant Project (GEF MGP) and Biodiversity Monitoring Survey (BMS)

This project was originally a 5 year initiative of the Government of Sri Lanka funded by the GEF, implemented by the UNDP, and executed by the Ministry of Environment & Natural Resources (MENR). The project started in late 2000 and continued till early 2006 and it was designed to target the conservation and sustainable use of globally significant biodiversity of a sub-set of Sri Lanka's lowland rain forests [17, 18]. The project activities were concentrated within Galle and Matara sections of Sinharaja and Kanneliya part of the KDN Complex in Galle District. The objective set for the above project was that:

> "The project will aim to secure the protection of the ecosystems in the rainforests of Sinharaja and Kanneliya-Dediyagala-Nakiyadeniya (KDN Complex) through community co-management". In order to achieve this overall objective, four clusters of outputs were set out:

a. Integrated buffer zone (Transition Zone according to the Biosphere Reserve Concept) community development focusing on biodiversity conservation and livelihood issues.
b. Strengthened institutional mechanisms to involve the community in decision-making.
c. Use of non-timber forest products on a sustainable basis.
d. Improve forest protection from encroachment and illicit logging.

The project report also states that:

> "these two forest sites are among the richest in terms of plant diversity and endemism within the whole country, and within their specific biome – the 'Lowland Rain Forests or wet forests'. Despite this ranking, the forests were quite intensively logged for hardwood timber in the 1970s to 1980s."

Furthermore, the Terminal Evaluation of the GEF MGP [17] has recommended "the consideration of a strong monitoring survey on biodiversity values, building on earlier documentations, e.g., NCR report [11] and previously conducted Medicinal Plant Surveys and Assessments etc. This became an excellent opportunity to ascertain the contribution of community co-management of the Transition Zones of these two International Biosphere Reserves (IBRs) towards their natural regeneration. Therefore, UNDP implemented the above recommendation in 2008 by supporting the present BMS project in order to determine the changes in the status of plant biodiversity in Sinharaja and Kanneliya forests since the NCR survey was conducted in 1992.

1.4. National Conservation Review (1991-1996)

The NCR was conducted mainly to address a major criticism of the first Forestry Master Plan for Sri Lanka (19) for its failure to address the environmental aspects of forestry particularly with respect to conserving the country's rapidly dwindling natural forests in the wet zone. The NCR undertook the task of evaluating all remaining natural forests (including mangroves) with respect to their importance for biodiversity, in terms of ecosystem and species diversity, and their value for soil and water conservation. Its over-riding objective was to define a national system of conservation forests in which watersheds important for soil conservation and hydrology are protected, forest biodiversity is fully represented and cultural, economic and social needs are met.

Legally designated forests, i.e., Forest Reserves (FR) and Proposed Forest Reserves (PR), and Other State Forests (OSF) of 200 ha or more were surveyed for biodiversity, except those in the north and east of the island which were then politically inaccessible. The biodiversity assessment was restricted to woody plants, vertebrates, mollusks and butterflies. Sampling of the sites was carried out by *gradsect sampling* (gradient-directed transect) method which involves a deliberate selection of transects which contain the steepest environmental gradients with maximum access present in an area. The gradsect sampling method was an efficient technique for rapidly assessing species diversity within natural forests. A total of 1,725 plots (100 m x 5 m) were inventoried along 310 gradsects in 204 forests. Analyses were based on approximately 69,400 records of 1,153 woody plant species and 24,000 records of 410 species of selected animal groups. Meanwhile a total of 281 forests were assessed, using existing data, with respect to their importance for soil and water conservation, e.g., soil erosion, headwaters protection, flood hazard and fog interception at higher altitudes. Despite some limitations, the NCR is among the most detailed, comprehensive and innovative evaluations of its kind carried out in any tropical country [11].

The results of the NCR project are contained in the *Environment Information Management System (EIMS)* database in the custody of the Forest Department. The EIMS database provides a very powerful tool that can be used to explore alternative options for biodiversity conservation, e.g., for generating optimum systems of conservation forest according to predefined criteria and quantifying their potential benefits and costs. Inevitably, designing an optimum system of conservation forests is an iterative process, particularly as more data become available as conservation forests are established and surveyed for management planning purposes. New data incorporated within EIMS can be used to refine conservation planning at the system level. Therefore, the Forest Department, in collaboration with other sectors, was encouraged to consider pragmatic ways of addressing the conservation priorities emanating from this study, taking into consideration their cultural, economic and social implications.

The NCR study clearly demonstrated the paramount importance of Sri Lanka's natural forests, both in terms of their role in maintaining ecosystem stability and functions and as a reservoir of high species diversity. It was also possible to identify those forests of highest importance for watershed protection and for biodiversity conservation at ecosystem and species levels. The close degree of complementary nature between the watershed and biodiversity values of forest highlighted the importance of such an integrated approach to conservation [11].

Although the NCR survey was conducted in western, eastern and northern aspects of Sinharaja in the Ratnapura District and southern parts in Galle and Matara Districts, the BMS was restricted only to the southern aspects as the activities of GEF MGP were located in the latter Districts. The forest areas in these southern aspects have not been logged, but they were encroached for slash and burn (chena) cultivation and tea in the past. In contrast, GEF MGP activities in KDN were located in the Kanneliya part in Galle District and the main forest degradation activity here has been the state-sponsored selective logging. Although the peripheral lands had also been encroached for chena (slash and burn) cultivation, such plots were not contained in the NCR sampled areas. The results have indicated that these two sites are the richest in terms of plant biodiversity in the Southern Province.

2. METHODS

The study areas for both NCR and BMS surveys were the same and they consisted of lowland rain forests. Some aspects of the study areas are shown in Figure 3 & 4. However, the time allocated for sampling various sites in NCR survey was an important factor that determined the sampling sizes used in various sites and therefore the sampling sizes used for Sinharaja and Kanneliya were different in both surveys (Table 2). As GEF MGP activities were confined to Galle and Matara Districts in Sinharaja and Kanneliya forest in Galle District respectively, the BMS was conducted along original NCR gradsects within the above areas. The survey method was exactly what was used for the NCR [11].

Figure 3. Aspects of the study area in Kanneliya. A = Study area: forest North of Kabbale Kanda. B = Aapproach from Opatha: forest in background was surveyed (Gradsect 34 plots R-V); C = Aapproach from Opatha: tea land bordering forest. D = BMS Field Team (left to right): Jayasuriya, Panduka, Abayawardana, Upali, Chitrasekera, Nihal, Wilbert and Nalin.

Figure 4. Aspects of the study area in Sinharaja. A = BMS field work in progress. B = Rainy & misty weather (note BMS paint marks on trees and undergrowth of bamboo). C = Dense undergrowth of bamboo (*Ochlandra stridula*).

2.1. Gradsect Sampling

Gradsect sampling was designed to provide a description of the full range of biodiversity within a region, overcoming the problems of inadequate representative sampling and accessibility, while minimizing survey costs. Transects were oriented along altitudinal gradients in order to sample the full range of biodiversity within a forest. Reference to the 1: 62,630 topographical maps enabled transects to be positioned at right angles to contours, ensuring that full range of altitudes and aspects were covered by one or more transects within each forest. Schematic diagram in Figure 5 shows how transects were aligned along altitudinal gradients, changing direction to maximize variability and taking advantage of available access routes. Plots were 100 m long, aligned along the length of the gradsect, and 5 m wide and each plot was spaced with at least 50 - 100 m gap before the start of the next plot (Figure 6). All woody species observed within each plot were inventoried (11).

Specimens of partially identified and unidentifiable species, as well as other species that are worthwhile recording, were collected and subsequently processed at the National Herbarium (NH). They were identified using relevant literature and reference specimens at the NH. As a reciprocal gesture, a set of mounted and fully labeled specimens were donated to the NH, as practiced during the NCR project that donated over 3,500 specimens to the NH.

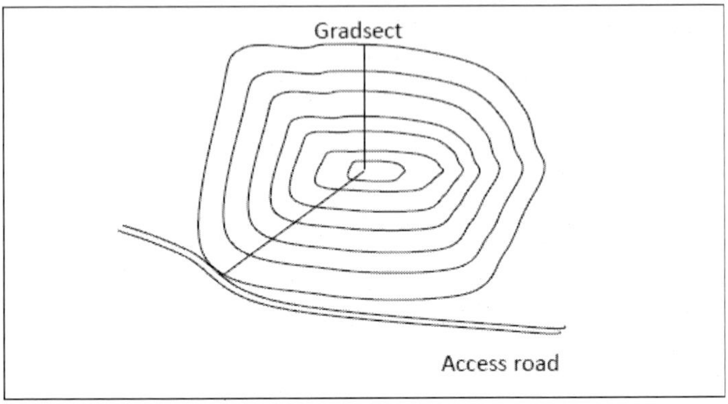

Figure 5. Schematic diagram showing alignment of a gradsect in relation to altitudinal gradient and access route.

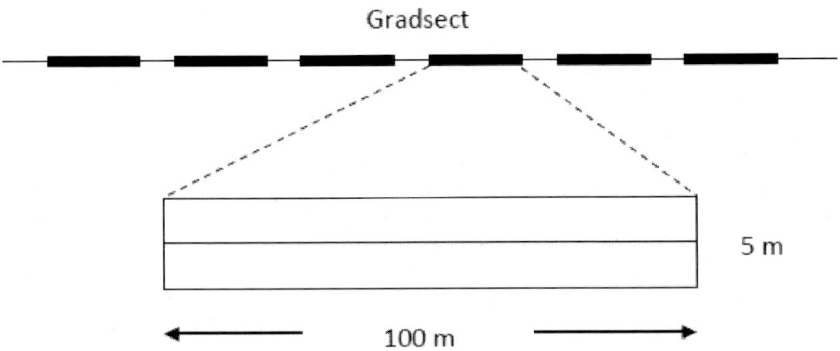

Figure 6. Schematic diagram of plots aligned along a gradsect at regular intervals.

2.2. Relocation of NCR Gradsects and Plots after Sixteen Years

The presence of three members of the original NCR field team (1992) in the BMS (2008) was an immense advantage in relocating NCR gradsects. GPS locations for all NCR gradsects were not available. The entry points of all gradsects in Kanneliya were accurately found in most areas while in some all original plots were precisely relocated (e.g., Kanneliya approached from Opatha). In one area, viz. in Kanneliya above Narangas Ella, two plot areas were seriously disturbed / displaced by a landslide that occurred in 2003. However, in Sinharaja, although the gradsects of the two surveys lay in close proximity to each other, they did not exactly overlay each other.

Figure 7. Relocation of NCR gradsects and plots in Kanneliya: A = NCR paint mark (1992) above, BMS mark being painted below (2008). B = NCR paint mark below (1992), BMS mark being painted above (2008). C = NCR paint mark (1992). D = NCR paint mark left (1992), BMS paint mark right (2008). E = Nylon loop in NCR plot (1992). F = NCR gradsect direction on rock (1992).

Prior to BMS, it was assumed that the markings of the original NCR gradsects and plots (yellow paint markings on tree trunks along plots and gaps and nylon loops on the branches of small trees at the beginning and end of each plot) would have been lost after over 16 years. However, it was an amazing find that these markings were found surviving in many plots in Kanneliya (Figure 7: A, B, C, D, E, F). GPS readings were noted, but often they failed as satellite signals were not received under high forest canopies.

2.3. Collection of Specimens

413 plant specimens were collected, mostly in 2 - 3 duplicates, from the two sites during the three field sessions. All specimens were identified and data used in the analysis. A complete set of specimens were processed, preserved and donated to the NH.

3. Results and Discussion

3.1. Protected Areas with High Plant Diversity (Hotspots)

In order to check biodiversity richness of Sinharaja and Kanneliya, it was useful to compare their plant diversity attributes with those of other biodiversity-rich protected areas. Information generated by the NCR survey would be a strong and reliable basis for this purpose [11]. Table 1 gives some important data for most biodiversity-rich protected areas in Sri Lanka.

It is evident that KDN complex harbors the richest plant biodiversity with highest number of endemic species in any protected area in the island. It is hypothesized that lowland rainforests located relatively at lower altitudes, e.g., KDN, are influenced by relatively high temperatures that facilitate higher rate of speciation and evolution, as indicated by relatively high quality biodiversity. If rainfall regimes remain within comparable ranges, such *lowland* rainforests show relatively more *lowland* character, compared with rainforests located at relatively higher altitudes, e.g., Sinharaja. This hypothesis is validated by above observations and however, it should be tested for various sites by analyzing a wider set of data, also in consideration of other factors such as sampling sizes, habitat diversity and disturbance levels etc.

The rich plant biodiversity in the KDN is remarkable since much of it has been logged for timber in the past, as also was highlighted in previous reports [13, 20]. This warrants enhanced attention on its conservation as a local and regional biodiversity hotspot. Furthermore it should be noted that detailed site comparison is not within the scope of this exercise.

Table 1. Woody plants in most biodiversity-rich protected areas in Sri Lanka

Site	Families	Genera	Species	Endemics
KDN Complex (excluding Derangala, Hindeneittu, Tawalama & Homadola)	?	?	358*	196*
[Kanneliya	61	147	263*	176*
Dediyagala	56	125	188	124
Nakiyadeniya]	59	161	236	132
Sinharaja (excluding Tibbotuwakota, Selvakanda & Kurulugala)	71	191	337*	192*
Knuckles	67	183	288	85
Gilimale-Eratne	70	179	279	164
Peak Wilderness	65	155	256	148
Morapitiya-Runakanda	63	166	244	131
Warathalgoda	61	155	217	126
Kalugala	65	157	216	116
Neluketiya	65	163	211	101
Mulatiyana	66	150	204	105
Dellawa	62	146	203	111

Sources: [13, 20].

* Woody plant species recorded in Kanneliya and Sinharaja outside NCR and BMS gradsect / plots were added.

3.2. Survey Extents for NCR and BMS at Sinharaja and Kanneliya

The extents of the two surveys, i.e., NCR (1992) and BMS (2008), are shown in Table 2.

Table 2. NCR Survey (1992) and BMS Survey (2008) of Sinharaja and Kanneliya forests

Attribute	Sinharaja		Kanneliya	
	NCR (1992)	BMS (2008)	NCR (1992)	BMS (2008)
Survey dates	27-28 Mar. 1992 9-10 May 1992 Field days = 4	20-22 Sep. 2008 21-23 Oct. 2008 Field days = 6	11-12 May 1992 14-18 Jul. 1992 Field days = 7	7-12 May 2008 17-18 Sep. 2008 Field days = 8
Gradsects	2 (Gradsects Nos. 27 & 33)	2 (Gradsect Nos. 27 & 33)	1 (Gradsect No. 34)	1 (Gradsect No. 34)
Plots	7 + 7 = 14	6 + 6 = 12	24	17
Sample size (ha)	0.70	0.60	1.20	0.85

3.3. Species Inventories

Species inventories of both surveys along with endemism, threatened status and status of ecological succession are given in Appendix 1. This enables the checking of presence/absence of a given species, genus or family in a given site or both sites. The summary of important attributes of species recorded in Sinharaja and Kanneliya are shown in Table 3. This is useful for site comparison or to cumulate information to synthesize a bigger picture of these most biodiversity- rich sites in Sri Lanka. Useful analyses can be performed using such data by EIMS program held by the Environmental Management Division of the Forest Department; e.g., defining an optimal system of forest sites for conservation using attributes such as endemism and rarity etc. [13, 20].

Table 3. Species recorded in Sinharaja and Kanneliya

Attribute	Sinharaja	Kanneliya	Both sites
NCR Survey (1992)	212	238	312
BMS (2008)	205	211	290
Both surveys	265	284	361
Endemics – NCR (percentage)	121 (57%)	153 (64.3%)	
Endemics – BMS (percentage)	111 (54.1%)	139 (66.0%)	
Total endemics (percentage)	145 (54.7%)	181 (63.7%)	214 (59.2%)

3.4. Changes in Biodiversity Attributes Since NCR Survey (1992)

Considerable changes in some important biodiversity attributes after sixteen years were noted during the BMS. The lower number of species recorded during the BMS in both sites was merely due to smaller sample sizes in the BMS (Table 3). In general, the relatively richer plant biodiversity at Kanneliya was evident from the number of species recorded in the two surveys.

3.4.1. Endemism

The proportional increase of endemic species by 1.7% in Kanneliya indicated the improvement of the biodiversity quality after 16 years (Table 3). It should be noted that in Kanneliya, gradsects and most plots in the two surveys overlaid with each other with high accuracy. However, in Sinharaja, the BMS showed a negative trend with 2.9% decrease of the number of endemics. This unexpected discrepancy was thought to be due to site variation between the two surveys, as none of the gradsects, and therefore the plots, of the BMS did not exactly overlay with those of the NCR surveys (see section 2.2).

Responses of endemic species to some types of forest disturbance have received the attention of some ecologists. Field observations and other studies in the selectively logged areas in Sinharaja [21] and partially disturbed rain forest in Kottawa in Galle District [22] showed the return of at least some of the canopy and sub-canopy endemic species as the forest recovers. What we know much less is the response of endemic species in the small life-form groups, i.e., how tolerant are they to disturbance, is not well known [2]. The regenerating vegetation in sites that were selectively logged, clear-felled for skid trails or abandoned after shifting cultivation was compared with that in an undisturbed part of the forest in Sinharaja [21]. This study reported that the population densities of 31 endemic species were distinctly lower in disturbed sites even after seven years of recovery. This has been viewed as a serious detrimental effect following such disturbances by Gunatilleke et al. [2].

3.4.2. Threatened Status of Species

A total of 346 species were evaluated for their conservation status at national and global levels according to the 2012 Red List [12]. Threatened status of each species recorded in the two sites is indicated in the Species Inventory (Appendix 1). A comparative and cumulative account of threatened status of plant species in the two sites is given in Table 4 and Figure 8.

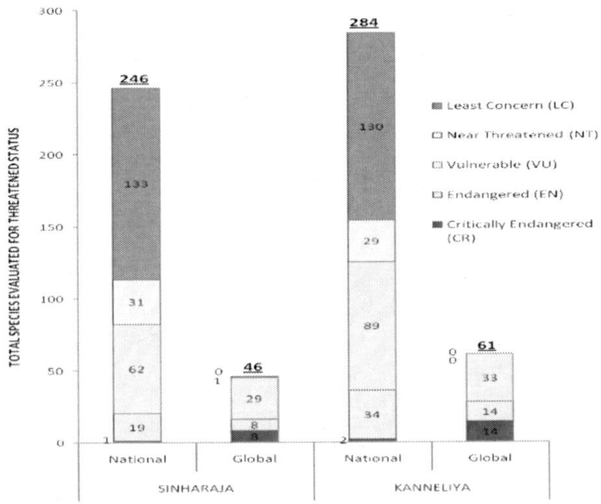

CR = Critically Endangered, EN= Endangered, VU = Vulnerable, NT = Near Threatened, LC = Least Concern.

Figure 8. Threatened status categories of species in Siharaja and Kanneliya.

Table 4. Threatened status categories of species in Sinharaja and Kanneliya

Site	Nationally threatened status							Globally threatened status						
	CR	EN	VU	NT	LC	Total threatened	%	CR	EN	VU	NT	LC	Total threatened	%
Sinharaja	1	19	62	31	133	113	45.9	8	8	29	0	1	45	18.3
Kanneliya	2	34	89	29	130	154	54.2	14	14	33	0	0	61	21.5
Both sites	2	45	97	51	151	195	66.4	16	18	53	1	2	88	25.5

Source: [12].

CR = Critically Endangered, EN= Endangered, VU = Vulnerable, NT = Near Threatened, LC = Least Concern.

Absolute values of threatened species should not be compared between the two sites as their sampling sizes differ and therefore the relative proportions of threatened plants can indicate a better comparison (Table 3). It is evident that Kanneliya harbors a higher proportion of nationally threatened species (54.2%) compared to that in Sinharaja (45.9%) and also a higher proportion of globally threatened species (21.5%) than that in Sinharaja (18.3%). These findings clearly indicated that Kanneliya functions as a relatively more important refugium for threatened plants.

In order to detect changes in the status of threatened species that occurred since the NCR survey was carried out in 1992, the numbers of nationally and globally threatened species recorded during the two surveys were compared (Table 5 and Figure 9).

In Kanneliya, the distinct increase in the proportion of both nationally (3.5%) and globally (0.3%) threatened species indicated that its biodiversity quality has markedly improved by the appearance of additional threatened species, thus rare species, during the recess of 16 years since the selective logging was ceased. However, the Sinharaja revealed degradation in biodiversity quality in that a reduction of nationally threatened (2.7%) and globally threatened species (4.2%) was indicated and it is difficult to explain this observation, except by the fact that the positions of the gradsects of the two surveys differed slightly.

Table 5. Changes in status of threatened species between NCR and BMS in Sinharaja and Kanneliya

Attributes of threatened status	**Sinharaja**		**Kanneliya**	
	NCR	BMS	NCR	BMS
Total number of species recorded	212	205	238	211
Number of nationally threatened species	107	98	126	119
Proportion of nationally threatened species (%)	50.5	47.8	52.9	56.4
Number of globally threatened species	47	37	66	59
Proportion of globally threatened species (%)	22.2	18.0	27.7	28.0

Source: [12].

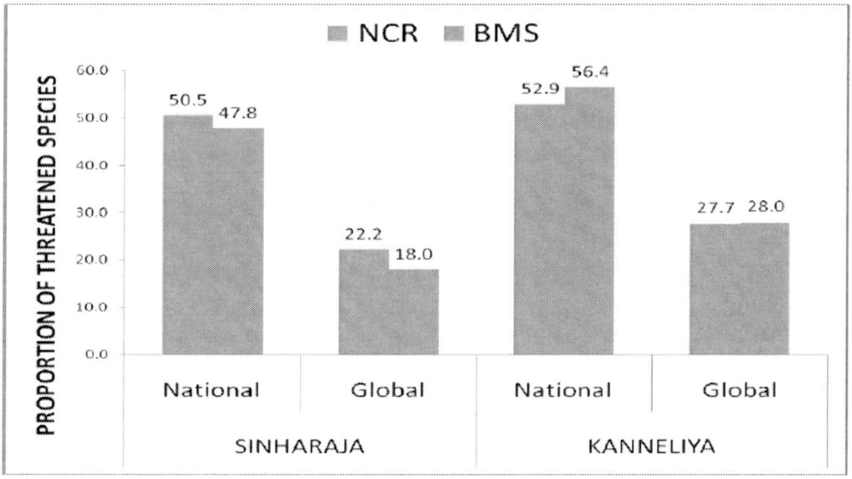

Source: [12].

Figure 9. Proportions of threatened species in Sinharaja and Kanneliya.

3.4.3. Succession Status and Recovery of Forest Profile and Biodiversity

Succession status is related to the rate of appearance of pioneer and secondary species in relation to primary species in a disturbed forest site that is in a state of recovery. Pioneers are highly light-tolerant, soft wooded, fast growing and possess the ability to produce large quantities of fruits and seeds and efficient dispersal mechanisms, e.g., *Macaranga peltata, Alstonia macrophylla* etc. These include some exotic weedy species that are often invasive, e.g., *Clidemia hirta*. Secondary species are often primary species that are relatively more light tolerant and fast-growing and succeed pioneers in a previously disturbed forest site. The secondary tree species are relatively more hard wooded than those of pioneer tree species and sometimes endemic, e.g., *Scumacheria castaneifolia and Dillenia triquetra*. In a satisfactorily regenerating site, primary species succeed secondary species and the rate of succession is highly influenced by the presence of forest seed source in the vicinity of the regenerating site. Conversely repeated disturbance, e.g., clearing, burning and cultivation, pushes back the recovery process.

Table 6. Succession status in Sinharaja and Kanneliya

Site	**Sinharaja**			**Kanneliya**		
	NCR (1992)	BMS (2008)	Change %	NCR (1992)	BMS (2008)	Change %
Total species	212	205		238	211	
Primary species (A) %	175 82.5%	179 87.3%	 +4.8%	203 85.3%	190 90.0%	 +4.7%
Secondary species (B) %	25 11.8%	17 8.2%	 -3.5%	27 11.3%	19 9.0%	 -2.3%
Pioneer species (C) %	12 5.7%	9 4.4%	 -1.3%	8 3.4%	2 1.0%	 -2.4%
Total Proportion of secondary and pioneer species %	17.5%	12.6%	-4.8%	14.6%	9.9%	-4.7%

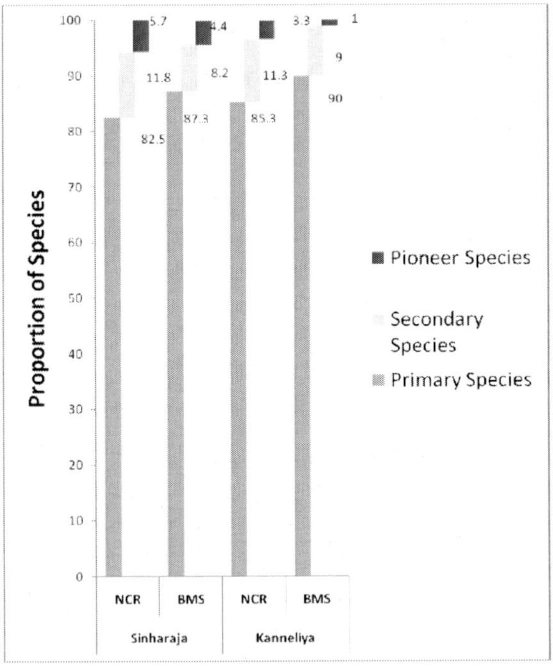

Figure 10. Succession status in Siharaja and Kanneliya.

Between the two surveys (NCR and BMS), both Sinharaja and Kanneliya showed an increase of primary species, i.e., +4.8% and +4.7% respectively, accompanied by a similar decrease of combined pioneer and secondary species, i.e., -4.8% and -4.7% respectively (Table 6 and Figure 10). This clearly indicated that forests at both sites are successfully regenerating and in Kanneliya the improvement in biodiversity quality is considerable in view of the recess that it was provided since selective logging.

3.4.3.1. Succession Status in an Abandoned Chena (Slash and Burn) Block

An abandoned chena plot along NCR / BMS gradsect No. 33 (Plot B) in Sinharaja showed considerably high rate of recovery as indicated by an 13.8% increase in the appearance of primary species accompanied by a similar decrease in combined pioneer and secondary species (Table 7 and Figure 11. Therefore, it is evident that the rate of succession or the recovery in highly degraded landscapes, such as slash and burn plots, the process is initially rapid, but becomes gradually slower with the progress of the recovery. No abandoned chena plots were encountered within the study areas in Kanneliya.

Table 7. Succession status in an abandoned chena plot in Sinharaja

Succession Group	NCR (1992) Plot 33 B	BMS (2008) Plot 33 B	Change %
Total species recorded	53	54	
Primary species (A) %	28 52.8%	36 66.6%	+13.8
Secondary species (B) %	19 35.8%	10 18.5%	-17.3
Pioneer species %	6 11.3%	8 14.8%	+3.5
Total number of secondary and pioneer species %	25 47.1%	18 33.3%	-13.8

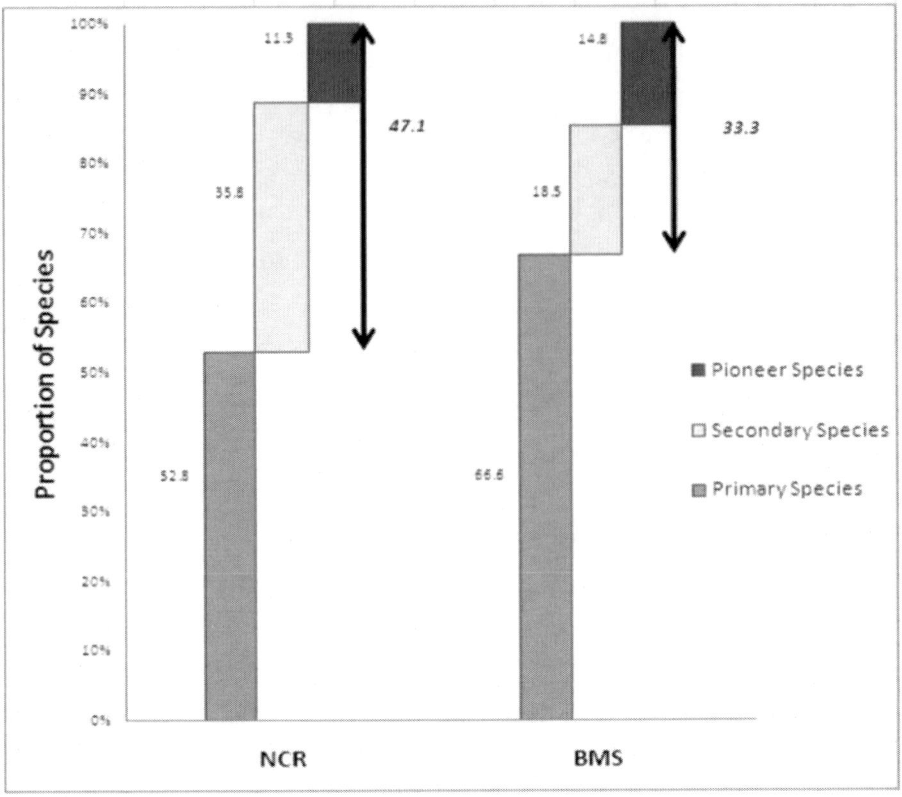

Figure 11. Succession status in an abandoned chena plot in Sinharaja.

3.4.4. Recovery of Forest Profile

In Kanneliya, majority of the plots surveyed for NCR was in logged areas where the forest profile was considerably disturbed and altered due to the loss of emergents and canopy trees and clearance for skid trails, timber trails and road network. Light regime among various strata too has been altered and at horizontal scale, a network of artificial light gaps have been created. Diversity and the rhythm of the forest seed production changed retrogressively. Such changes paved the way for the establishment of secondary species from within the forest and that of pioneer species, usually from outside the forest. The result was the forest degradation in terms of stature, composition and biodiversity quality.

The enforcement of a moratorium on logging in 1990, finally granted "ecological freedom" to Kanneliya. After nearly two decades of recess, observations indicated recovery of forest stature accompanied by the appearance of emergents (trees above canopy level) and reproductively mature canopy trees gradually bringing back the original seed bank into the forest dynamics. Spreading canopies ensured the stabilization of the light regime and soil and water conservation (Figure 12).

Figure 12. Kanneliya: A = Forest recovering from selective logging. B = Forest canopy in recovering forest;

Study site in Sinharaja has not been selectively logged. However, peripheral influences such as illicit logging, firewood and other non-timber forest products (NTFP) gathering and clearance for chena cultivation have occurred in the recent and distant past. Estimation of disturbance in both surveys clearly indicated disturbance in peripheral plots (edge effect) and relatively more pristine condition of the interior plots.

3.4.5. Recovery in Timber and Skid Trails

In Kanneliya timber and skid trails have been made to facilitate logging. Although such timber and skid trails were eluded from NCR and BMS surveys, forest re-growth clearly showed saplings of many primary species growing fast in them along with some secondary species. Paucity of the pioneers within majority of these trails was also evident. Apparently, the presence of the forest on either side of these trails has provided a close-range seed source, shade, nutrients and other favorable conditions for speedy recovery.

3.5. Enrichment Planting in Rain Forests

After the selective logging ceased in Kanneliya in about 1990, enrichment planting with Mahogany (*Swietenia macrophylla*) was carried out using line-planting technique. Mahogany is a light-demanding tree which does not establish and grow properly under the shade of the forest. The line-planting technique involved the marking of a parallel series of lines that were separated by 10m (33 feet) gaps starting from a baseline and these planting lines required clearing of forest strips of several meters wide to provide space and light for the growth of Mahogany seedlings planted at 10m intervals (Figure 13). Thus a large number of seedlings and saplings of primary forest tree species and other shrubs had to be removed to prepare the land for line planting, inflicting more damage to already logged forest. Some information obtained from Range Forest Office, Tawalama is presented in Table 8.

Table 8. Enrichment planting in Kanneliya with Mahogany

Year	Forest Beat	Unit	Extent (Ha)	Timber Species
1989	Hiniduma	Nugagala	20	Mahogany
1989	Opatha	Halgahawela	20	Mahogany
1991	Udugama	Koralegama	80	Mahogany
1996	Opatha	Weerapana	50	Hora and Mahogany

Source: Range Forest Office, Tawalama (September, 2008).

A = Line clearing for Mahogany planting; undergrowth removed to receive sufficient light – 13 May 1992. B = Line clearing for Mahogany planting, undergrowth removed to receive sufficient light – 14 July 1992.

Figure 13. Enrichment planting in Kanneliya.

First segment of the NCR gradsect No. 34 (Plots A-G) crossed this "forest enrichment" area and Mahogany saplings were recorded in it in 1992. However the BMS in 2008 noted that Mahogany has disappeared along with their enrichment lines which have been reclaimed by the regenerating natural forest. Mahogany apparently survived only along roadsides where sufficient light was available. This feature is also exhibited by the appearance and disappearance of *Alstonia macrophylla* in forest gaps, e.g., timber trails and openings, regulated by the availability of light. Therefore the enrichment planting with exotic species in selectively logged Lowland Rain Forests was a proven failure that led not only to further damage of the regenerating forest but a waste of financial and other resources. Experience (personal communication with Messrs. N. H. Chitrasekera and Martin Wijesinghe) has also shown that Mahogany seedlings/saplings are devoured by herbivores such as porcupines when planted inside the forest. Sinhakumara [15] also has recommended that the enrichment planting with Mahogany be abandoned in Kanneliya forest.

3.5.1. Correct Use of the Enrichment Planting Technique

Enrichment planting with selected tree species is an effective technique that can be used to improve and accelerate timber production and environment qualities in degraded forests. However its applicability and success is greatly guided by the condition of the land and the choice of species; otherwise it will lead to failure inflicting further damage to the enrichment area and loss of resources, e.g., line planting with Mahogany in Kanneliya forest.

Enrichment should never be done in selectively logged forests where a large number of individuals that remain belong to primary status. Such species provide a source of propagules, mainly seeds, necessary for the recovery process. However, it could be done with success in relatively more degraded and open plots (e.g., recovering chena lands) with a considerable number of individuals belonging to pioneer and secondary succession status and if such degraded plots have a reasonable amount of tree cover for the establishment of the planted saplings. A mixture of indigenous and exotic species that are generally fast-growing and light-tolerant such as *Dipterocarpus zeylanicus*, *Swietenia macrophylla*, *Michelia champaca*, *Cananga odorata*, *Cyathocalyx zeylanica*, *Alstonia macrophylla* and species of rattan, e.g., *Calamus zeylanicus, C. ovoideus* is recommended for this purpose. A mixed species concept is expected to be at an advantage in re-colonization of a degraded forest habitat rather than the use of a single species. In highly open plots such as abandoned chena lands and fern lands, initial conditioning of the landscape with suitable pioneer species as nurse plant cover will be necessary for the establishment of the succeeding enrichment species.

3.6. Extraction of Non-Timber Forest Products (NTFP)

It was observed that in Kanneliya, extraction of NTFP by the bordering communities was limited to resins obtained from many species of Dipterocarpaceae (Figure 14). The products are sold to village shop owners who sell them to traders in Colombo. Only a few individuals (in Koralegama) were found to be engaged in resin collection and even they do this only as a subsidiary income, in keeping with the tradition, as they own plots of tea and get reasonable income. Resin collectors, invariably very skilled tree climbers, generally visit the same resin-producing trees on routine basis (e.g., monthly). Although many *Dipterocarpus glandulosus*

('Dorana'*)* trees that were tapped for 'Dorana' oil were found in the forest, this practice which finally kills the tree is apparently vanishing in Kanneliya (Figure 14). The most important NTFP are food products (treackle – liquid palm sugar and jaggery – crystalline palm sugar) of the 'Kitul' tree (*Caryota urens*). Although in the past 'Kitul' trees in the forest were tapped for sugar sap, at present the trees that grow in homesteads and edges of forest that are not far from the village are used for tapping. In general, the dependence on NTFP has decreased in the recent past.

Rogers & Abayawardana [17] states that there is no demand for fuel wood or poles; these are immediately available in home gardens. The only NTFP in past use is specialist food (fruits or *Caryota* tapping for treackle) and medicinal plants; but *Caryota* is increasingly planted in home gardens and a wide variety of medicinal plants are also available in home gardens.

However, a new disastrous practice of collecting wild cinnamon bark has recently developed around Kanneliya, and perhaps in other areas too. The bark of all species of *Cinnamomum* is collected in large scale to manufacture popular medicinal application (balm) and this practice is organized by commercial entrepreneurs engaging villagers to collect bark from the wild.

A = Collection of resin from *Doona affinis (Dipterocarpaceae) tree;* B = *Dipterocarpus glandulosus* (Dipterocarpaceae) tree tapped for oil.

Figure 14. NTFP collection in Kanneliya.

CONCLUSION

Influence of GEF MGP on the Conservation of Target Sites

Operation of GEF MGP during 2000 to 2006 apparently had a positive influence on the conservation of these forest sites and the results of the BMS has clearly indicated that GEF MGP has achieved its objective, i.e., "*to secure the protection of the ecosystems in the rainforests of Sinharaja and KDN complex through community co-management.*"

ACKNOWLEDGMENTS

We wish to thank UNDP for supporting this project and Forest Department for entrusting us with the task of conducting this project. Dedicated support in the field was provided by Messrs. Panduka Weerasinghe, Program Assistant and Daminda Sumanapala, Development Assistant of the Forest Department. Excellent support was provided by Mr. Priyantha Seneviratne, Divisional Forest Officer (DFO), Galle, Rashal Weerasinghe, DFO, Matara, Range Forest Officers in Tawalama, Neluwa, Deniyaya and Sinharaja Forest Ranges, relevant Beat Forest Officers and Field Assistants in making arrangements to accommodate the field team during three field sessions and providing guidance and assistance in the field. Late Mr. Y. R. Premasiri, Additional RFO, Deniyaya who provided us with support in the field at Sinharaja is remembered with great honor. Mr P. P. Sirisena, Retired BFO's guidance at Sinharaja was critical in locating the area that was surveyed over sixteen years. Mr. N. H. Chitrasekera of Koralegama provided wonderful support as a field guide, a resource person with his wealth of local knowledge, and a companion in the field and during discussions held in the evenings after concluding field work each day.

The wholehearted and energetic support of Mr. Upali Weerasekera, Herbarium Assistant, National Herbarium, was crucial both in the field and at the herbarium.

We also thank Dr. Siril Wijesundara, Director General, National Botanic Garden for providing all necessary facilities at the National Herbarium, Peradeniya to identify and process plant specimens collected in the field. This support was provided by Mr. D. H. P. Peramunugama, Director, Ms. Achala Attanayake (Deputy Director), Ms. R. A. S. W. Ranasinghe, Research Officer, Ms. N. P. T. Gunawardana, Agriculture Monitoring Officer, Ms. F. A. Padmini, Laboratory Assistant, and Ms. Kumaramenike, Herbarium Assistant.

The assistance given by Mr. Rasika Jayasuriya in the preparation of graphics is thankfully acknowledged while our sincere thanks are due to the staff of the EML Consultants and the Integrated Development Association (IDEA) for their unstinted support and offering us a very pleasant and comfortable environment to conduct this work.

RERERENCES

[1] De Zoysa, N. and R. Raheem 1987. *Sinharaja – A Rainforest in Sri Lanka*, March for Conservation, Sri Lanka. 92.

[2] Gunatilleke, C. V. S., I. A. U. N. Gunatlleke, E. U. K. Ethugala and S. Esufali 2004.*Ecology of Sinharaja Rain Forest and the Forest Dynamics Plot in Sri Lanka's Natural World Heritage Site*. WHT Publications (Pvt.) Ltd., Sri Lanka.

[3] Jayasinghe, C. D., S. P. Ekanayake, C. N. B. Bambaradeniya, S. Amarasinghe. 2005. *Tropical Rainforests*: A Resource Book for Secondary School Teachers. IUCN, Sri Lanka pp. 144.

[4] Banyard, S. G. and D. Fernando, 1988. Sinharaja Forest: monitoring changes by using aerial photographs of two different dates. *The Sri Lanka Forester* 18: 101-108.

[5] Forest Department, Ministry of Lands and Land Development, Sri Lanka and WWF/IUCN. Sri Lanka. 1988. *Project proposal – Sinharaja conservation project.*

[6] Forest Department, Ministry of Lands and Land Development, and WWF/IUCN, Sri Lanka, 1993. *Management plan for the conservation of the Sinharaja forest – phase II.* pp. 92.

[7] IUCN. *Management plan for the conservation of the Sinharaja forest.* 1993. Forest Department, Sri Lanka (unpublished).

[8] Bandaratillake, H. M. 1992. *Managing the buffer zone in Sinharaja World Heritage Forest. Parks.* A report prepared for the Forest Department, Sri Lanka.

[9] Dassanayake, M. D., F. R. Fosberg, and W. D. Clayton (Editors). 1980-2000. *Revised Handbook to the Flora of Ceylon*, Vol. 1- 14. Amerind, New Delhi.

[10] Jayasuriya, A. H. M. and S. Balasubramaniam. 1990. *Flora of Sinharaja.* Unpublished Report, Natural Resources, Energy & Science Authority (NARESA), Sri Lanka.

[11] IUCN and WCMC. 1997. *Designing an optimum protected areas system for Sri Lanka's natural forests, Vol 1 & 2.* Forest Department, Ministry of Forestry and environment, Battaramulla, Sri Lanka, pp. 299. (IUCN and WCMC. 1997, 60-70).

[12] Ministry of Environment. 2012. *The National Red List 2012 of Sri Lanka: Conservation status of the Fauna and Flora.* Ministry of Environment, Colombo, Sri Lanka 476 pp.

[13] Nisbet, R. H. McD. 1961. *A Management Inventory of the Kanneliya-Nakiyadeniya—Dediyagala Forest Area, Ceylon.* Canada-Ceylon Colombo Plan Project, 1961.

[14] Peeris, C. V. S. 1975. *The ecology of the endemic tree species of Sri Lanka in relation to their conservation*. PhD Thesis, University of Aberdeen, UK.

[15] Sinhakumara, B. M. P. 1994. *Ecological Assessment of KDN Forest Complex*. Report prepared for IUCN, Sri Lanka and Forest Department of the Ministry of Agriculture, Land and Forestry,Sri Lanka, 74 pp.

[16] Ashton, P. S. and C. V. S. Gunatilleke. 1987. New light on the plant geography of Ceylon I. Historical plant geography. *Journal of Biogeography* 14:249-285.

[17] Rogers, A. and S. D. *Abayawardana. Contributing to the Conservation of Unique Biodiversity in the Threatened Rain Forests of South-west Sri Lanka.* Terminal Evaluation Project Report prepared for Forest Department, Government of Sri Lanka, GEF & UNDP, Colombo.

[18] Bandaratillake, H. M. 2008. *Contributing to the Conservation of Unique Biodiversity in the Threatened Rain Forests of South-west Sri Lanka.* Terminal Project Report prepared for Forest Department. Government of Sri Lanka, GEF & UNDP, Colombo.

[19] Forestry Planning Unit, Ministry of Agriculture, Lands & Forestry, Sri Lanka. 1995. *Sri Lanka Forestry Sector Master Plan.* Battaramulla, Sri Lanka pp. 32-85.

[20] IUCN. 1993. *Conservation Review of some Natural Forests in Sri Lanka.* Report prepared for the Environmental Management in Forestry Development, Forest Department, Sri Lanka, pp. 163.

[21] De Zoysa, N, C. V. S. Gunatilleke, C. V. S. and I. A. U. N. Gunatilleke. 1990. *Comparative phytosociology of natural and modified rain forest sites in Sinharaja MAB reserve in Sri Lanka. Rain forest Regeneration and Management.* Man & the Biosphere Series, Vol. 6 (Eds. Gomez Pompa, A., T. C. Whitmore and M. Hadley). UNESCO, Paris and The Parthenon Publishing Group Ltd., Carnforth, UK., pp. 215-224.

[22] Gunatilleke, C. V. S. and P. S. Ashton. 1987. New light on the plant geography of Ceylon, II. Ecological biogeography of the lowland endemic tree flora. *J. Biogeography* 14: 295-327.

[23] Angiosperm Phylogeny Group. 2009. An update of the Angiosperm Phylogeny Group classification for the orders and families of flowering plants: APG III, *Botanical Journal of the Linnean Society,* 161 (2): 105–121.

APPENDIX 1: SPECIES INVENTORY OF KANNELIYA AND SINHARAJA FORESTS

Table A-1.

KAN-NCR (1992): Kanneliya NCR Survey (1992)
KAN-BMS (2008): Kanneliya Biodiversity Monitoring Survey (2008)
SIN-NCR (1992): Sinharaja NCR Survey (1992)
SIN-BMS (2008): Sinharaja Biodiversity Monitoring Survey (2008)
E: Endemic to Sri Lanka
THR-NAT: Threatened Status – National [12]
CR = Critically Endangered; EN = Endangered; VU = Vulnerable
THR-GLOB: Threatened Status – Global [12]
CR = Critically Endangered; EN = Endangered; VU = Vulnerable;
NT = Near Threatened; LC = Least Concerned
SUC-Status: Succession Status: A = Primary species; B =Secondary species; C = Pioneer species, including plantation species
The plant classification and nomenclature is basically according to (9), and updated according to (23).

Summary

Table A-2.

Sinharaja	**Kanneliya**
NCR SURVEY (1992) = 212 species	NCR SURVEY (1992) = 238 species
BMS (2008) = 205	BMS (2008) = 211
TOTAL NO. OF SPECIES = 265	TOTAL NO. OF SPECIES = 284
ENDEMIC SPECIES – NCR = 121	ENDEMIC SPECIES – NCR = 153
ENDEMIC SPECIES – BMS = 111	ENDEMIC SPECIES – BMS = 139
TOTAL ENDEMIC SPECIES = 145	TOTAL ENDEMIC SPECIES = 181
NATIONALLY THREATENED SPECIES = 113	NATIONALLY THREATENED SPECIES = 154
GLOBALLY THREATENED SPECIES = 45	GLOBALLY THREATENED SPECIES = 61

Both Sites

Table A-3.

NCR SURVEY (1992) = 312 species
BMS (2008) = 290
TOTAL NO. OF SPECIES = 361
ENDEMIC SPECIES = 214
EVALUATED FOR THREATENED STATUS = 346
NATIONALLY THREATENED SPECIES = 195
GLOBALLY THREATENED SPECIES = 88

Table A-4.

FAMILY **Species** **(Synonym)**	**KAN-NCR (1992) Grad-sects = 1 Plots = 24**	**KAN-BMS (2008) Grad-sects = 1 Plots = 17**	**SIN-NCR (1992) Grad-sects = 2 Plots = 14**	**SIN-BMS (2008) Grad-sects = 2 Plots = 12**	**E**	**THR-NAT**	**THR-GLOB**	**SUC-Status**
ACANTHACEAE								
Strobilanthes lupulina (S. asperrima)			*	*		LC		A
ACHARIACEAE (FLACOURTIACEAE)								
Hydnocarpus octandra	*	*	*	*	E	LC	VU	A
Trichadenia zeylanica	*		*	*	E	LC	VU	A
ANACARDIACEAE								
Camnosperma zeylanicum	*	*	*	*	E	LC		A
Mangifera zeylanica	*	*	*	*	E	LC	VU	A
Nothopegia beddomei			*	*		LC		A
Semecarpus acuminata		*	*	*	E	VU	EN	A
Semecarpus gardneri	*	*	*	*	E	LC	VU	A
Semecarpus marginata	*	*	*	*	E	NT	VU	A
Semecarpus moonii	*	*			E	VU	VU	A
Semecarpus parvifolia		*		*	E	VU	VU	A
Semecarpus pubescens	*		*		E	VU	VU	A
Semecarpus subpeltata	*	*		*	E	VU	VU	A
Semecarpus walkeri	*		*	*	E	LC	VU	A

FAMILY **Species** **(Synonym)**	**KAN-NCR (1992) Grad-sects = 1 Plots = 24**	**KAN-BMS (2008) Grad-sects = 1 Plots = 17**	**SIN-NCR (1992) Grad-sects = 2 Plots = 14**	**SIN-BMS (2008) Grad-sects = 2 Plots = 12**	**E**	**THR-NAT**	**THR-GLOB**	**SUC-Status**
ANISOPHYLLACEAE (RHIZOPHORACEAE)								
Anisophyllea cinnamomoides	*	*	*	*	E	NT		A
ANNONACEAE								
Artabotrys zeylanicus	*	*	*	*		LC		A
Cyathocalys zeylanica		*	*	*		LC		A
Desmos elegans	*	*	*		E	VU		A
Desmos zeylanica			*		E	NT		A
Enicosanthum acuminata	*	*	*	*	E	LC		A
Goniothalamus gardneri	*				E	VU	EN	A
Goniothalamus hookeri	*	*	*		E	VU	VU	A
Goniothalamus salicina ssp. reticulata	*	*			E	VU		A
Goniothalamus salicina ssp. salicina	*	*	*	*	E	VU		A
Goniothalamus thomsonii		*		*	E	VU		A
Goniothalamus thwaitesii				*		NT		A
Phoenicanthus obliqua	*	*	*	*	E	NT		A
Polyalthia korinti	*	*				LC		A
Polyalthia percicaefolia			*		E	EN		A

Table A-4. (Continued)

FAMILY **Species** **(Synonym)**	**KAN-NCR (1992) Grad-sects = 1 Plots = 24**	**KAN-BMS (2008) Grad-sects = 1 Plots = 17**	**SIN-NCR (1992) Grad-sects = 2 Plots = 14**	**SIN-BMS (2008) Grad-sects = 2 Plots = 12**	**E**	**THR-NAT**	**THR-GLOB**	**SUC-Status**
Uvaria cordata		*				VU		A
Uvaria semecarpifolia	*	*	*	*	E	LC		B
Uvaria sphenocarpa	*	*	*	*	E	LC		A
Xylopia championii	*	*	*	*	E	LC		A
Xylopia parvifolia			*			LC		A
APOCYNACEAE								
Alstonia macrophylla	*		*	*				C
Alstonia scholaris				*		LC	LC	C
Carissa spinarum	*	*				LC		A
Cleghornia acuminata	*		*	*	E	VU		A
Hunteria zeylanica	*	*	*	*		NT		A
Pagiantha dichotoma			*	*		LC		B
Petchia ceylanica	*	*			E	NT		A
Willughbeia cirrhifera	*	*			E	VU	VU	A
ARACEAE								
Pothos parvispadix		*		*	E	EN		A
ARALIACEAE								
Schefflera emarginata			*	*	E	VU		A

FAMILY Species (Synonym)	KAN-NCR (1992) Grad-sects = 1 Plots = 24	KAN-BMS (2008) Grad-sects = 1 Plots = 17	SIN-NCR (1992) Grad-sects = 2 Plots = 14	SIN-BMS (2008) Grad-sects = 2 Plots = 12	E	THR-NAT	THR-GLOB	SUC-Status
ARECACEAE								
Calamus delicatulus	*	*	*	*	E	VUEN		A
Calamus digitatus	*	*		*	E	VU		A
Calamus ovoideus				*	E	EN		A
Calamus pseudotenuis	*	*	*	*		VU		A
Calamus rivalis	*	*	*	*	E	VU		A
Calamus zeylanicus	*	*	*	*	E	EN		A
Caryota urens	*	*	*	*		LC		B
Loxococcus rupicola	*				E	EN	CR	A
Oncosperma fasciculatum	*			*	E	VU	NT	A
ARISTOLOCHIACEAE								
Thottea siliquosa (Apama siliquosa)	*	*	*	*		LC		A
ASPARAGACEAE (DRACAENACEAE, LILIACEAE)								
Asparagus falcatus		*	*	*		LC		B
Dracaena thwaitesii	*	*	*	*		NT		A
BURSERACEAE								
Canarium zeylanicum	*	*	*	*	E	VU	VU	A
BUXACEAE								
Sarcococca zeylanica			*		E	VU		A

Table A-4. (Continued)

FAMILY Species (Synonym)	KAN- NCR (1992) Grad- sects = 1 Plots = 24	KAN- BMS (2008) Grad- sects = 1 Plots = 17	SIN- NCR (1992) Grad- sects = 2 Plots = 14	SIN- BMS (2008) Grad- sects = 2 Plots = 12	E	THR- NAT	THR- GLOB	SUC- Status
CALOPHYLLACEAE (CLUSIACEAE)								
Calophyllum acidus	*	*	*	*	E	NT		A
Calophyllum bracteatum	*	*	*	*	E	NT	VU	A
Calophyllum moonii	*	*	*	*	E	VU	VU	A
Calophyllum thwaitesii	*	*	*	*	E	VU	VU	A
Mesua ferrea		*				LC		A
Mesua pulchella	*	*			E			A
Mesua thwaitesii			*	*	E	LC		A
CANNABACEAE (ULMACEAE)								
Trema orientale			*					C
CELASTRACEAE								
Euonymus walkeri	*		*		E	LC	VU	A
Kokoona zeylanica	*				E	EN		A
Loeseneriella arnottiana	*	*		*		EN		A
Microtropis wallichiana	*	*	*	*		LC		A
Salacia diandra	*	*	*	*	E	EN		A
Salacia reticulata			*	*		EN		A

FAMILY **Species** **(Synonym)**	**KAN-NCR (1992) Grad-sects = 1 Plots = 24**	**KAN-BMS (2008) Grad-sects = 1 Plots = 17**	**SIN-NCR (1992) Grad-sects = 2 Plots = 14**	**SIN-BMS (2008) Grad-sects = 2 Plots = 12**	**E**	**THR-NAT**	**THR-GLOB**	**SUC-Status**
CENTROPLACACEAE (CELASTRACEAE)								
Bhesa ceylanica	*	*	*	*	E	LC		A
Bhesa nitidissima		*			E	LC		A
CLUSIACEAE								
Garcinia echinocarpa	*		*	*		VU		A
Garcinia hermonii	*	*	*	*	E	VU	LR/cd	A
Garcinia morella	*	*		*		NT		A
Garcinia quaesita	*	*	*	*	E	LC	VU	A
Garcinia terpnophylla	*	*	*		E	EN	VU	A
Garcinia thwaitesii		*	*		E	EN	EN	A
COMBRETACEAE								
Terminalia zeylanica (T. parviflora)	*		*	*		LC	VU	A
CONNARACEAE								
Connarus championii	*	*	*	*	E	NT		A
Rourea minor	*	*	*	*		LC		A
CONVOLVCULACEAE								
Argyreia thwaitesii		*			E	LC		B
Erycibe paniculata	*	*	*			LC		A

Table A-4. (Continued)

FAMILY **Species** **(Synonym)**	**KAN-NCR (1992) Grad-sects = 1 Plots = 24**	**KAN-BMS (2008) Grad-sects = 1 Plots = 17**	**SIN-NCR (1992) Grad-sects = 2 Plots = 14**	**SIN-BMS (2008) Grad-sects = 2 Plots = 12**	**E**	**THR-NAT**	**THR-GLOB**	**SUC-Status**
CORNACEAE (ALANGIACEAE)								
Alangium salviifolium ssp. hexapetalum	*					NT		A
Mastixia macrophylla	*					VU	VU	A
Mastixia tetrandra	*	*	*	*		LC		A
CRYPTERONIACEAE (MELASTOMATACEAE)								
Axinandra zeylanica	*	*	*		E	VU		A
DICHAPETALACEAE								
Dichapetalum gelonioides	*					LC		A
Dichapetalum zeylanicum (D. helferianum)			*	*	E	NT		A
DILLENIACEAE								
Dillenia retusa	*	*				LC		A
Dillenia triquetra (Wormia triquetra)	*	*	*	*		LC		B
Schumacheria angustifolia	*				E	EN		A
Schumacheria castaneifolia	*	*	*	*	E	LC		B
Tetracera sarmentosa			*			LC		B

FAMILY **Species** **(Synonym)**	**KAN-** **NCR** **(1992)** **Grad-** **sects = 1** **Plots = 24**	**KAN-** **BMS** **(2008)** **Grad-** **sects = 1** **Plots = 17**	**SIN-** **NCR** **(1992)** **Grad-** **sects = 2** **Plots = 14**	**SIN-** **BMS** **(2008)** **Grad-** **sects = 2** **Plots = 12**	**E**	**THR-** **NAT**	**THR-** **GLOB**	**SUC-** **Status**
DIPTEROCARPACEAE								
Dipterocapus glandulosus		*			E	EN	CR	A
Dipterocarpus hispidus	*	*			E	VU	CR	A
Dipterocarpus insignis	*	*			E	EN	CR	A
Dipterocarpus zeylanicus		*			E	NT	EN	A
Doona affinis	*	*	*	*	E	VU	EN	A
Doona congestiflora	*	*	*	*	E	VU		A
Doona gardneri			*		E	VU	CR	A
Doona macrophylla	*	*		*	E	VU	CR	A
Doona nervosa	*	*	*	*	E	VU	CR	A
Doona oblonga	*	*		*	E	VU	EN	A
Doona trapezifolia	*	*	*	*	E	VU	CR	A
Doona venulosa	*				E	VU	EN	A
Hopea discolor			*	*	E	EN	EN	A
Hopea jucunda	*	*			E	VU		A
Hopea modesta	*	*			E	EN		A
Shorea dyeri	*				E	VU		A
Shorea hulanidda	*	*			E	EN		A
Shorea lissophylla		*	*		E	VU	CR	A
Shorea oblongifolia	*	*	*		E	VU	CR	A

Table A-4. (Continued)

FAMILY **Species** **(Synonym)**	**KAN-NCR (1992) Grad-sects = 1 Plots = 24**	**KAN-BMS (2008) Grad-sects = 1 Plots = 17**	**SIN-NCR (1992) Grad-sects = 2 Plots = 14**	**SIN-BMS (2008) Grad-sects = 2 Plots = 12**	**E**	**THR-NAT**	**THR-GLOB**	**SUC-Status**
Shorea pallescens	*				E	EN	CR	A
Shorea stipularis	*	*			E	VU	CR	A
Stemonoporus acuminatus			*	*	E	EN	EN	A
Stemonoporus bullatus		*			E	CR	EN	A
Stemonoporus canaliculatus			*		E	EN	CR	A
Sunaptea scabriusculum (Cotylelobium scabriusculum)	*	*			E	EN		A
Vateria copallifera	*		*		E	VU	EN	A
Vatica affinis	*	*			E	EN	CR	A
EBENACEAE								
Diospyros acuminata	*	*	*	*	E	VU	VU	A
Diospyros acuta	*	*			E	EN	EN	A
Diospyros albiflora	*	*			E	EN	VU	A
Dispyros chaetocarpa	*	*			E	EN	VU	A
Diospyros hirsuta	*	*	*	*	E	VU	VU	A
Dispyros insignis	*	*	*	*		LC		A
Diospros insignis var. parvifolia	*	*		*				A
Diospros oblongifolia	*	*	*		E	VU	VU	A

FAMILY Species (Synonym)	KAN- NCR (1992) Grad- sects = 1 Plots = 24	KAN- BMS (2008) Grad- sects = 1 Plots = 17	SIN- NCR (1992) Grad- sects = 2 Plots = 14	SIN- BMS (2008) Grad- sects = 2 Plots = 12	E	THR- NAT	THR- GLOB	SUC- Status
Diospyros oppositifolia	*				E	EN	EN	A
Diospyros pemadasai		*			E	EN		A
Diospyros quaesita	*	*			E	EN	VU	A
Diospyros racemosa			*	*		VU		A
Diospyros ? sylvatica		*				VU		A
Diospyros thwaitesii	*	*	*		E	VU	VU	A
Diospyros walkeri	*					VU	VU	A
ELAEOCARPACEAE								
Elaeocarpus amoenus		*			E	VU		A
Elaeocarpus glandulifer	*			*	E	VU	VU	A
Elaeocarpus serratus	*		*			LC		B
Elaeocarpus subvillosus	*	*	*	*	E	NT		B
ERYTHROXYLACEAE								
Erythroxylum lanceolatum				*		VU		A
Erythroxylum moonii			*	*		NT		A
EUPHORBIACEAE								
Agrostistachys coriacea	*		*	*	E	LC	VU	A
Agrostistachys hookeri		*			E	LC	CR	A
Agrostistachys intramarginalis (= A. cerumica)	*	*			E	LC		A
Chaetocarpus castanocarpus	*	*	*	*		LC		A

Table A-4. (Continued)

FAMILY Species (Synonym)	KAN- NCR (1992) Grad- sects = 1 Plots = 24	KAN- BMS (2008) Grad- sects = 1 Plots = 17	SIN- NCR (1992) Grad- sects = 2 Plots = 14	SIN- BMS (2008) Grad- sects = 2 Plots = 12	E	THR- NAT	THR- GLOB	SUC- Status
Chaetocarpus coriaceus	*			*	E		VU	A
Croton lacciferus			*	*		LC		B
Croton moonii	*	*			E	EN		A
Fahrenheitia minor	*	*			E	LC	CR	A
Fahrenheitia zeylanica	*	*	*	*		LC		A
Macaranga peltata	*		*	*		LC		C
Mallotus fuscescens	*	*	*		E	LC	VU	A
Mallotus rhamnifolius	*					VU		B
Ptychopyxis thwaitesii	*	*	*	*	E	VU		A
FABACEAE (LEGUMINOSAE)								
Adenanthera bicolor		*	*		E	NT		B
Caesalpinia bonduc	*					LC		B
Dalbergia pseudo-sisso	*	*	*	*		LC		A
Derris canarensis			*			NT		B
Entada pusaetha				*		LC		A
Humboldia laurifolia	*	*	*	*		LC	VU	A
GENTIANCEAE (LOGANIACEAE)								
Fagraea ceilanica	*	*	*	*		NT		A

FAMILY **Species** **(Synonym)**	**KAN-** **NCR** **(1992)** **Grad-** **sects = 1** **Plots = 24**	**KAN-** **BMS** **(2008)** **Grad-** **sects = 1** **Plots = 17**	**SIN-** **NCR** **(1992)** **Grad-** **sects = 2** **Plots = 14**	**SIN-** **BMS** **(2008)** **Grad-** **sects = 2** **Plots = 12**	**E**	**THR-** **NAT**	**THR-** **GLOB**	**SUC-** **Status**
ICACINACEAE								
Nothapodytes nimmoniana (N. foetida)	*		*			NT		A
LAMIACEAE (VERBENACEAE)								
Clerodendron infortunatum	*		*	*		LC		C
Vitex altissima	*	*	*	*		NT		A
LAURACEAE								
Actinodaphne albifrons	*	*	*	*	E	VU	VU	A
Actinodaphne candolleana	*	*			E	NT		A
Actinodaphne elegans			*	*	E	LC		A
Alseodaphne semecarpifolia	*		*			VU		A
Cinnamomum capparu-cornde		*			E	VU	VU	A
Cinnamomum dubium	*	*	*	*	E	VU		A
Cinnamomum sinharajaense			*	*	E	EN		A
Cinnamomum zeylanicum (Cinnamomum verum)		*		*		VU		A
Cryptocarya wightiana	*	*	*	*		NT	VU	A
Litsea gardneri	*	*	*		E	VU	VU	A

Table A-4. (Continued)

FAMILY Species (Synonym)	KAN-NCR (1992) Grad-sects = 1 Plots = 24	KAN-BMS (2008) Grad-sects = 1 Plots = 17	SIN-NCR (1992) Grad-sects = 2 Plots = 14	SIN-BMS (2008) Grad-sects = 2 Plots = 12	E	THR-NAT	THR-GLOB	SUC-Status
Litsea iteodaphne			*	*	E	VU	VU	A
Litsea longifolia	*	*	*	*	E	LC	VU	B
Litsea nemoralis		*			E		EN	A
Litsea sp.				*				A
Neotitsea cassia	*	*	*	*		LC		A
Persea macrantha			*	*		VU		A
LINACEAE								
Hugonia ferruginea			*			VU		A
LOGANIACEAE								
Strychnos wallichiana (S. cinnamonifolia)	*	*	*	*		NT		A
MALPIGHIACEAE								
Hiptage benghalensis	*					LC		B
MALVACEAE (TILIACEAE)								
Bombax ceiba			*			LC		B
Cullenia ceylanica	*	*	*	*	E	LC	VU	A
Cullenia rosayroana	*	*	*	*	E	LC	LR/cd	A
Grewia carpinifolia	*					LC		B

FAMILY Species (Synonym)	**KAN-NCR (1992) Grad-sects = 1 Plots = 24**	**KAN-BMS (2008) Grad-sects = 1 Plots = 17**	**SIN-NCR (1992) Grad-sects = 2 Plots = 14**	**SIN-BMS (2008) Grad-sects = 2 Plots = 12**	**E**	**THR-NAT**	**THR-GLOB**	**SUC-Status**
MELASTOMATACEAE (MEMECYLACEAE)								
Lijndenia capitellata	*	*	*	*	E	VU	CR	A
Melastoma malabathricum	*	*		*		LC		C
Memecylon clarkeanum	*		*	*	E	NT	VU	A
Memecylon discolor				*	E	VU	EN	A
Memecylon giganteum		*		*	E	EN	EN	A
Memecylon grande		*			E	ENEN	VU	A
Memecylon macrocarpum	*				E	EN		A
Memecylon procerum	*	*		*	E	VU		A
Memecylon rivulare	*	*			E	VU		A
Memecylon rostratum	*	*	*	*	E	NT		A
Memecylon royenii	*	*			E	LC		A
Memecylon sylvaticum			*		E	NT		A
Memecylon varians	*	*	*	*		VU		A
MELIACEAE								
Agiaia apiocarpa	*	*	*	*		LC	VU	A
Dysoxylum ? ficiforme	*		*			NT		A

Table A-4. (Continued)

FAMILY Species (Synonym)	KAN-NCR (1992) Grad-sects = 1 Plots = 24	KAN-BMS (2008) Grad-sects = 1 Plots = 17	SIN-NCR (1992) Grad-sects = 2 Plots = 14	SIN-BMS (2008) Grad-sects = 2 Plots = 12	E	THR-NAT	THR-GLOB	SUC-Status
Swietenia macrophylla	*							C
Walsura trifoliolata ssp. acuminata		*		*		LC		A
MENISPERMACEAE								
Coscinium fenerstratum	*	*	*	*		LC		A
MONIMIACEAE								
Hortonia angustifolia	*		*		E	VU	CR	A
Hortonia floribunda				*	E	EN		A
MORACEAE								
Artocarpus nobilis	*	*	*	*	E	LC	VU	B
Ficus diversiformis	*	*			E	LC		A
Ficus fergusoni	*		*	*	E	LC		A
Ficus hispida			*			LC		B
Ficus laevis var. laevis		*				LC		A
MUSACEAE								
Musa balbisiana			*	*		EN		A
MYRISTICACEAE								
Horsfieldia irya	*		*	*		LC		B
Myristica dactyloides	*	*	*	*		LC	LC	A

FAMILY Species (Synonym)	KAN-NCR (1992) Grad-sects = 1 Plots = 24	KAN-BMS (2008) Grad-sects = 1 Plots = 17	SIN-NCR (1992) Grad-sects = 2 Plots = 14	SIN-BMS (2008) Grad-sects = 2 Plots = 12	E	THR-NAT	THR-GLOB	SUC-Status
MYRTACEAE								
Cleistocalyx operculatus (C. nervosum)			*	*		LC		A
Eugenia glabra		*			E	EN	EN	A
Eugenia rivolurum	*		*		E	VU	VU	A
Eugenia rufo-fulva		*			E	EN	VU	A
Eugenia terpnophylla		*			E	VU	EN	A
Syzygium alubo (S. makul)	*	*	*	*		NT		A
Syzygium amphoraecarpus (S. aqueum)		*		*	E	NT		A
Syzygium caryophyllatum			*			LC	EN	B
Syzygium cordifolium	*				E	VU		A
Syzygium cylindricum	*		*	*	E	LC		A
Syzygium firmum	*	*	*	*	E	LC		A
Syzygium gardneri		*		*		LC		A
Syzygium lewisii				*	E	VU		A
Syzygium lissophyllum	*		*	*	E	NT		A
Syzygium micranthum			*	*	E	LC		A
Syzygium neesianum	*	*		*		LC	VU	A

Table A-4. (Continued)

FAMILY Species (Synonym)	KAN-NCR (1992) Grad-sects = 1 Plots = 24	KAN-BMS (2008) Grad-sects = 1 Plots = 17	SIN-NCR (1992) Grad-sects = 2 Plots = 14	SIN-BMS (2008) Grad-sects = 2 Plots = 12	E	THR-NAT	THR-GLOB	SUC-Status
Syzygium sp.		*						A
Syzygium wightianum (S. lanceolatum)		*		*		VU		A
NEPENTHACEAE								
Nepenthes distillatoria	*	*	*	*	E	LC	VU	B
OCHNACEAE								
Gomphia serrata	*	*	*	*		LC		A
OLACACEAE								
Strombosia nana	*	*	*		E	VU		A
OLEACEAE								
Chionanthus albidiflora	*				E	VU	CR	A
PANDANACEAE								
Freycinetia walkeri	*	*	*	*	E	NT		A
Pandanus thwaitesii		*	*	*		NT		A
PASSIFLORACEAE								
Adenia hondala	*			*		LC		A
PENTAPHYLLACEAE (THEACEAE)								
Adinandra lasiopetala		*		*	E	EN		A
Eurya acuminata			*	*		NT		B

FAMILY Species (Synonym)	KAN- NCR (1992) Grad- sects = 1 Plots = 24	KAN- BMS (2008) Grad- sects = 1 Plots = 17	SIN- NCR (1992) Grad- sects = 2 Plots = 14	SIN- BMS (2008) Grad- sects = 2 Plots = 12	E	THR- NAT	THR- GLOB	SUC- Status
PHYLLANTHACEAE (EUPHORBIACEAE)								
Antidesma bunius			*			LC		A
Antidesma walkeri	*		*		E	LC		A
Aporusa cardiosperma	*	*	*	*		LC	VU	B
Aporusa lanceolata	*	*	*	*	E	LC	VU	A
Breynia vitis-idaea			*			LC		B
Bridelia moonii	*	*			E	LC	VU	
Cleistanthus pallidus	*				E	LC		A
Cleistanthus patulus	*	*	*					A
Cleistanthus robustus				*	E	VU	CR	A
Glochidion acutifolium				*	E	NT		A
Glochidion moonii			*	*	E	LC		A
Glochidion nemorale	*	*	*	*	E	LC		A
Glochidion stellatum	*	*		*	E	LC		A
Glochidion zeylanicum		*		*		LC		B
Suregada angustifolia	*					LC		A
PIPERACEAE								

Table A-4. (Continued)

FAMILY Species (Synonym)	KAN-NCR (1992) Grad-sects = 1 Plots = 24	KAN-BMS (2008) Grad-sects = 1 Plots = 17	SIN-NCR (1992) Grad-sects = 2 Plots = 14	SIN-BMS (2008) Grad-sects = 2 Plots = 12	E	THR-NAT	THR-GLOB	SUC-Status
Piper trineuron	*			*	E	NT		A
POACEAE								
Ochlandra stridula	*	*	*	*	E	LC		B
POLYGALACEAE								
Xanthophyllum zeylanicum (X. geminiflorum)	*	*			E	LC		A
PRIMULACEAE (MYRSINACEAE)								
Ardisia gardneri	*	*			E	LC		A
Ardisia lankaensis		*			E	VU		A
Ardisia moonii	*		*	*	E	LC		A
Ardisia pauciflora			*	*		NT		A
Maesa indica (M. perrottetiana)			*	*		LC		C
RHAMNACEAE								
Ventilago gamblei	*	*		*		LC		A
Zizyphus napeca			*		E	LC		A
RHIZOPHORACEAE								
Carallia brachiata		*	*	*		NT		A
Carallia calycina	*	*			E	VU	VU	A

FAMILY Species (Synonym)	**KAN-NCR (1992) Grad-sects = 1 Plots = 24**	**KAN-BMS (2008) Grad-sects = 1 Plots = 17**	**SIN-NCR (1992) Grad-sects = 2 Plots = 14**	**SIN-BMS (2008) Grad-sects = 2 Plots = 12**	**E**	**THR-NAT**	**THR-GLOB**	**SUC-Status**
ROSACEAE								
Prunus walkeri			*	*	E	LC		A
RUBIACEAE								
Aidia gardneri	*				E	VU		A
Byrsophyllum ellipticum	*	*		*	E	VU		A
Canthium campanulatum			*	*	E	NT		A
Canthium rheedii			*			NT		A
Chassalia curviflora var. ophioxyloides			*			LC		A
Dichilanthe zeylanica	*				E	VU	EN	A
Gaertnera divaricata	*	*			E	VU		A
Gaertnera rosea	*		*		E	LC	VU	A
Gaertnera vaginans	*	*	*	*		LC		B
Gaertnera sp.	*							A
Gardenia crameri	*				E	VU		A
Hedyotis fruticosa			*			LC		C
Hedyotis sp.			*					C
Ixora jucunda	*	*	*	*	E	LC	VU	A
Lasianthus moonii	*	*	*	*	E	LC		A
Lasianthus obliquus	*	*	*	*	E	LC		A
Lasianthus oliganthus	*		*	*	E	LC		A

Table A-4. (Continued)

FAMILY Species (Synonym)	KAN- NCR (1992) Grad- sects = 1 Plots = 24	KAN- BMS (2008) Grad- sects = 1 Plots = 17	SIN- NCR (1992) Grad- sects = 2 Plots = 14	SIN- BMS (2008) Grad- sects = 2 Plots = 12	E	THR- NAT	THR- GLOB	SUC- Status
Lasianthus strigosus				*	E	LC		A
Mussaenda samana (M. glabrata)	*				E	LC		C
Nargedia macrocarpa	*	*	*	*	E	LC		A
Neurocalyx calycinus				*		VU		A
Pavetta blanda			*		E	LC		A
Pavetta indica	*		*			LC		B
Psilanthus travancorensis (Coffea travancorensis)			*	*		VU		A
Psychotria dubia			*		E	NT	VU	A
Psychotria glandulifera	*				E	VU	EN	A
Psychotria nigra var. nigra	*	*	*	*		LC		A
Psychotria sarmentosa	*	*	*	*		NT		A
Psychotria waasii		*			E	NT	VU	A
Psydrax dicoccus (Canthium dicocca)	*		*			LC		A
Saprosma foetens ssp. ceylanicum				*	E	LC		A
Timonius flavescens (T. jambosella)	*	*	*	*		LC		B
Uncaria elliptica	*	*	*	*		LC		B
Urophyllum ellipticum	*		*	*	E	LC		A

FAMILY **Species** **(Synonym)**	**KAN-NCR (1992) Grad-sects = 1 Plots = 24**	**KAN-BMS (2008) Grad-sects = 1 Plots = 17**	**SIN-NCR (1992) Grad-sects = 2 Plots = 14**	**SIN-BMS (2008) Grad-sects = 2 Plots = 12**	**E**	**THR-NAT**	**THR-GLOB**	**SUC-Status**
Wendlandia bicuspidata			*		E	LC		C
RUTACEAE								
Acronychia pedunculata			*			LC		B
Clausena dentata	*					LC		B
Clausena indica	*					LC		B
Glycosmis cyanocarpa var. simplicifolia		*				EN		A
Glycosmis pentaphylla			*	*		LC		A
Luvunga angustifolia	*	*	*	*	E	LC		A
Melicope lunu-ankenda (Euodia lunu-ankenda)	*					LC		B
Toddalia asiatica			*	*		LC		C
SALICACEAE (FLACORTIACEAE)								
Casearia zeylanica	*	*				LC		A
Homalium ceylanicum	*	*		*		LC		A
Scolopia acuminata	*	*	*	*		LC		A
SAPINDACEAE								
Allophylus zeylanicus	*	*	*	*	E	LC	CR	A
Dimocarpus longan	*	*	*	*		LC		A

Table A-4. (Continued)

FAMILY **Species** **(Synonym)**	**KAN-NCR (1992) Grad-sects = 1 Plots = 24**	**KAN-BMS (2008) Grad-sects = 1 Plots = 17**	**SIN-NCR (1992) Grad-sects = 2 Plots = 14**	**SIN-BMS (2008) Grad-sects = 2 Plots = 12**	**E**	**THR-NAT**	**THR-GLOB**	**SUC-Status**
Filicium decipiens				*		LC		A
Lepisanthes erecta	*		*			VU		A
Pometia pinnata (P. tomentosa)	*		*	*		LC		A
SAPOTACEAE								
Chrysophyllum roxburghii	*	*				NT		A
Isonandra lanceolata	*	*	*	*		VU		A
Isonandra zeylanica	*	*	*	*	E	VU		A
Madhuca fulva	*	*	*		E	VU	VU	A
Mimusops elengi			*			NT		A
Palaquium canaliculatum		*			E	VU	EN	A
Palaquium grande	*	*	*	*	E	VU	VU	A
Palaquium hinmolpedda			*		E	VU		A
Palaquium pauciflorum	*	*		*	E	EN	VU	A
Palaquium petiolare	*	*	*	*	E	VU	LC	A
Palaquium rubiginosum	*	*			E	VU	VU	A
Palaquium thwaitesii		*			E	VU	VU	A
SCHIZANDRACEAE								
Kadsura heteroclita			*	*		EN		A

FAMILY Species (Synonym)	KAN- NCR (1992) Grad- sects = 1 Plots = 24	KAN- BMS (2008) Grad- sects = 1 Plots = 17	SIN- NCR (1992) Grad- sects = 2 Plots = 14	SIN- BMS (2008) Grad- sects = 2 Plots = 12	E	THR- NAT	THR- GLOB	SUC- Status
SIMAROUBACEAE								
Quassia amara (Samadara indica)	*	*				VU		A
SMILACACEAE								
Smilax zeylanica	*		*	*		LC		B
STAPHYLEACEAE								
Turpinia malabarica			*			LC		B
STEMONURACEAE (ICACINACEAE)								
Gomphandra tetrandra (Stemonurus tetrandra)	*	*	*	*		NT		A
Stemonurus apicalis (Urandra apicalis)	*	*	*	*	E	NT	CR	A
SYMPLOCACEAE								
Symplocos cochinchinensis	*		*	*		LC		C
Symplocos coronata			*	*	E	EN		A
Symplocos cuneata	*	*		*	E	EN		A
Symplocos diversifolia				*	E	CR		A
Symplocos pulchra var. hispidula	*		*	*	E	EN	VU	A
THEACEAE								
Camellia sinensis				*				C
THYMELAECEAE								

Table A-4. (Continued)

FAMILY Species (Synonym)	KAN- NCR (1992) Grad- sects = 1 Plots = 24	KAN- BMS (2008) Grad- sects = 1 Plots = 17	SIN- NCR (1992) Grad- sects = 2 Plots = 14	SIN- BMS (2008) Grad- sects = 2 Plots = 12	E	THR- NAT	THR- GLOB	SUC- Status
Gyrinops walla	*	*	*	*		VU		A
VITACEAE (LEEACEAE)								
Cissus latifolia (C. acuminata)	*	*	*	*		LC		A
Cissus lonchiphylla	*	*			E	NT		A
Leea indica	*	*	*			LC		B
XANTHORRHOEACEAE (PHORMIACEAE)								
Dianella ensifolia			*	*		LC		A
INDET (specimen Jayasuriya 9787)		*						A

In: India
Editors: J. Lessard and O. Rivière
ISBN: 978-1-53618-413-6

Chapter 3

AN OVERVIEW OF THE THREATS TO FAUNAL DIVERSITY OF RAJASTHAN

Neisseril Anirudhan Kashmeera and Ambalaparambil Vasu Sudhikumar*
Centre for Animal Taxonomy and Ecology, Department of Zoology,
Christ College, Irinjalakuda, Kerala, India

ABSTRACT

Being a desert state, the climate of Rajasthan is unique characterized by low rainfall, dryness and extreme temperature. Fauna of this region are specific, adapted to this hostile environment. This unique diversity of species makes Rajasthan a regional priority for conservation. Recently the exclusive fauna of this region is facing many unprecedented threats. Nowadays, the most serious threat is climate change. Oil exploration, overgrazing, mining of minerals and sandstones, habitat alteration and destruction caused by Indira Gandhi NaharPariyojna canals, etc. are the other prevailing threats. Current knowledge on the major threats to fauna of Rajasthan is briefly reviewed here. According to World Wildlife Fund India, this area has lost half of its wildlife in past forty years. So studies on threats to fauna of Rajasthan will assume increasing relevance as

*Corresponding Author's E-mail: avsudhi@christcollegeijk.edu.in.

government sectors seek to protect and conserve declining biodiversity. More interdisciplinary studies are required to warn about unnoticed threats to biodiversity of Rajasthan.

INTRODUCTION

Inspite of its hostile environment, Rajasthan is fairly rich in faunal life. Almost all major phyla of invertebrates and vertebrates ranging from protozoa to mammals are reported from the state of Rajasthan. Fauna of Rajasthan have been documented since time immemorial. A total of 82 species of protozoa belonging to three classes, Mastigophora, Sarcodina and Ciliata were reported from Rajasthan (Mahajan 1963, 1971, 1977). Nematodes of Rajasthan has been widely studied by Baqri and Bohra (2001), Bohra and Baqri (2003, 2005), Bohra and Sultana (2008, 2010, 2012). Tikader (1966) reported 13 species of spiders belonging to 6 families from desert area of Rajasthan. Later Sivaperuman and Rathore (2004) reported 28 species of spiders from Desert National Park. 24 species of ticks (Sanyal and De, 2005), 45 species of crustaceans (Biswas, 1971) were reported from Rajasthan. Khanna (2005) recorded six species of scolopendrid centipedes from Rajasthan. Sharma (2015) has worked out 47 species of Odonata fauna of Rajasthan. Termites have been extensively studied by Roonwal (1970-1980), Roonwal and Bose (1964, 1969), Roonwal and Rathore (1974, 1975) Roonwal and Sen-Sarma (1960), Roonwal and Verma (1977), Rathore (1986, 1987, 1989, 1994). Chhotani and Ray (1976), Tak (1995, 2008), Tak and Rathore (1996) have studied Hymenopteran fauna of Rajasthan.Vazirani (1970) has studied aquatic beetles and Sewak (2009 a, b) have studied dung beetles. The molluscs of the Rajasthan have been studied by Ray and Mukerjee (1969), Gupta (1973). The fishes have been extensively studied by Datta and Majumdar (1970). Biswas and Sanyal (1977) studied Reptilian fauna. Numerous works have been reported on birds of Rajasthan and some of the latest works include Chhangani (2017), Bhardwaj (2018), Kidwai&Bhattacharjee (2016), Yadav & Chauhan (2018). Mammals have been studied mainly by

Krishna and Prakash (1955, 1956, 1960), Agrawal (1967), Gupta and Agrawal (1966.1968), Biswas and Ghose (1970), Mohnot (1977, 1981), Sinha (1973, 1975, 1976, 1980), Prakash (1956, 1958, 1962, 1963 and 1977), Kankanae (2000).

Rajasthan is having rich biodiversity with four National Parks viz., Desert National Park, Keoladeo Ghana NP, Ranthambhore NP and Sariska NP. The State also has 25 Wildlife sanctuaries and 11 conservation reserves. However fauna of the state is facing several threats due to climate change, industrialization, ecotransformation due to Indira Gandhi NaharPariyojna (IGNP) Canals, Pollution, etc.

INDUSTRIALIZATION

Industrial sector in Rajasthan comprises cotton cloth, cement, sugar, urea, zinc ingots, ball bearings, edible oils, etc. Due to rapid industrialization, heavy metal contamination from industrial effluents also has increased. Zooplankton population has been effected by heavy metal toxicity from effluents of dyes and printing industries in Jodhpur (Verma and choudary, 2017). Heavy metal overdose can cause neurological complications, hypertension, liver and kidney dysfunctions, hematological disorders, etc. (Krishna and Govil, 2004). Studies on mammalian scat samples from Talchaper Blackbuck sanctuary and Dhavadoli protected area showed contamination with heavy metals like Pb, Cd, Cr, Cu and Zn (Gupta and Bakre, 2013). Contamination by heavy metals like Cd, Cr, Cu and Zn was also reported in Bikaner Zoological Garden (Gupta, 2013). Studies conducted on mammalian scat samples of Desert National Park, Jaisalmer and Gajner Wildlife sanctuary, Bikaner also showed contamination by heavy metals Pb, Cd, Cr, Cu and Zn (Gupta, 2012).

Fluoride toxicity is prevalent in various parts of Rajasthan due to high fluoride content in industrial chemical effluents. Chronic fluoride intoxication is characterized by mottling of teeth, osteosclerosis of skeleton and damage to nervous system in humans and animals (Choubisa, 2001). Fluorosis has been reported in 21 villages of Banswara, Dungarpur and

Udaipur districts. Fluorosis was reported in Dromedary camels in Rajasthan. Affected camels showed abortion, irregular estrus cycle, still births, dental and skeletal fluorosis (Choubisa, 2010a). Fluorosis was also observed in domestic horses and donkeys in Rajasthan (Choubisa, 2010b).

Industrial effluents can alter physical, chemical and biological nature of water bodies and thereby affect the survival of aquatic life. Studies conducted in Kota barrage showed that it is highly influenced by pollution. Benthic species diversity was considerably affected and was dominated by pollution indicator species like *Lymnaeaacuminata* and *Chironomus*larvae (Sarang and Sharma, 2014). Kishorsagarlake in Kota is dominated by diatom population designating that water quality of it is deteriorating (Kumar et al., 2008). Massive death of fishes has also been reported due to water pollution by industrial effluents in Chandralohiriver, Kota (Sultana et al., 2012).

Indira Gandhi NaharPariyojna (IGNP) Canals

IGNP is a gigantic venture aiming to transform desert wasteland into agriculturally productive area. IGNP canals bring water from Himalayas to vast stretches of arid western Rajasthan. In order to prevent the shifting sand dunes alongside the canal from blocking the flow of canal water, Government of Rajasthan has conducted afforestation programmes on both sides of canals. Studies show that these plantations provide shelter for various mammals and birds (Sivaperuman and Baqri, 2009). Besides these, crop patterns of this region have changed which attracted new agricultural pests. Also several minor pests have replaced major pests. Due to introduction of water, many grasslands were converted to croplands thus reducing the habitat of many native birds, although the number of water birds increased. Birds purely dependent on grasslands like Stoliczka'sbushchat, Great Indian Bustard, Houbara Bustard, white bellied sand grouse are facing severe threat due to this transformation (Sivaperuman and Baqri, 2009; Idris et al., 2009).

With increased availability of water and vegetation cover in arid areas, animals like Bandicoots, Nilgai and Wild boars are appearing in IGNP area (Sharma et al., 2002). Due to ecotransformation of desert by IGNP fruit plants are flourishing in this region resulting in emergence of previously unrecorded fruit bats (Purohit et al., 2020). Even soil fauna is also changing due to transformation of arid grasslands. Studies have reported increasing diversity of collembolan in Thar Desert of Sriganganagar district of Rajasthan (Faisal and Ahmed, 2005).

MINING

Pollution due to mining can harmfully affect health, survival and activities of humans and other living things (Vyas and Pancholi, 2009). Operation of various construction equipments and construction work itself in mines generate significant noise pollution. Blasting in mines generate vibration and noise which scare the wild fauna of nearby forest patches and force them to migrate to other areas. In operation phase of mines there is emission of toxic gases which have serious impact on terrestrial and aquatic animals. Migratory routes of birds will be affected due to change in topography, construction and blasting activities. There are chances of human-wildlife conflict due to habitat destruction (Sharma, 2015). Studies conducted in Ajmer showed that population of 22 species of birds declined due to mining (Meena et al., 2018). In Bijolia mining area rare species reported earlier like tigers, panthers, chinkaras, cheetals, hares, bluebulls, etc. disappeared due to mining activity (Chauhan, 2010). Mining activities at central parts of Aravalli mountain ranges has caused decline in the population of *Hemidactylus triedrus* (Sharma et al., 2014).

CLIMATE CHANGE

The Inter-Governmental Panel on Climate Change projected for hotter days and warmer nights and reduced rainfall in Thar Desert region by 21st

century. Such changes in climate will result in significant changes in biodiversity with changing life cycles, loss, migration and invasion of new habitat in Thar region (Rao et al., 2012).

OVERGRAZING

According to nineteenth livestock census 2012, total livestock population in Rajasthan is 577.32 lacs which constitutes 11 percent of the total livestock population in India. Livestocks in Rajasthan includes cattle, buffalo, sheep, goat, horse and ponies, mules, donkeys, camel, pig, etc. Pastures constitute only 5 percent of total geographic area of Rajasthan. Livestock keepers traditionally relied on common grazing lands "gochars," sacred groves "orans" and forests. Due to absence of management efforts overgrazing is resulting in the reduction of grasslands for grazing. Pastures are shrinking also due to growth of mining industry. Due to constant grazing pastures cannot recover resulting in their shrinking. Many animals are fully dependent on grasslands for their survival. Birds like Lesser Florican, Great Indian Bustard are purely grassland species. Animals like wild buffalo, barasingha, blackbuck, wild ass and chinkara are dependent on grasslands. The health of grasslands highly affects the survival of these animals.

INVASIVE ALIEN SPECIES

Thar Desert of Rajasthan has been severely invaded by *Prosopis juliflora, Lantana camara, Parthenium hysterophorus, Ageratum conyzoides, Argemone Mexicana, etc.* causing threat to survival of native biodiversity of the desert. The aquatic ecosystems are invaded by *Typha angustata, Ipomoea carnea, Eichhorniacrassipes, Hydrilla verticillata and Nelumbo nucifera* (Charan and Singh, 2018). As the indigenous plants are disappearing, the indigenous fauna dependent on them are also seriously affected.

Thar Desert of Rajasthan was once considered as unstable zone for mosquito breeding due to hot and arid climatic conditions. But changes in human population, water availability and increase in plantations in the environment have contributed to increase in breeding of mosquito in the desert (Singh et al., 2013).

Conclusion

Lion share of geographic area of Rajasthan is the Thar Desert, which being a fragile ecosystem, is further threatened because of the rapid increase in human population. Recently, human population in Rajasthan increased by 21.3%. Population growth also resulted in rapid urbanization and industrialization in this state which also became threat to the indigenous fauna. The rapid increase in the population of livestock has also resulted in degradation of the grasslands which are key habitat of several native species. Though Indira Gandhi Canal has transformed desert grassland into a fertile land, the indigenous biodiversity is under threat because the canal has changed the soil moisture, texture and vegetation composition. The major long-term threat faced by Rajasthan state is climate change. Changes in rainfall patterns, temperature, etc. severely affect the survival of native fauna.

References

Agrawal, V.C. (1967). New mammal records from Rajasthan. *Journal of Science and Technology*, 5: 342-344.

Baqri, Q.H. and Bohra, P. (2001). Nematodes from Rajasthan, India. I. Six new species of Dorylaimida. *Nematology*, 3 (2): 113-127.

Bhardwaj, D. K. (2018). Checklist of birds of JamwaRamgarh wild life sanctuary, district Jaipur, *Rajasthan.International Educational Applied Scientific Research Journal*, 3(5): 10-17.

Biswas, S. (1971). *Fauna of Rajasthan, India.* Part 11. Crustacea: Cladocera. Records of Zoological Survey of India, 63 (1965): 95-141.

Biswas, B. and Ghose, R.K. (1970). *Taxonomic notes on the Indian Pale Hedgehogs of the genus ParaechinusTrouessart with descriptions of a new species and subspecies*. Mammalia, Paris, 34 (8): 467-477.

Biswas, S. and Sanyal, D.P. (1977). Fauna of Rajasthan. India. Part: Reptilia. *Records of Zoological Survey of India, Delhi*. 73 (1975): 247-269.

Bohra, P. and Baqri, Q.H. (2003). Nematodes from Rajasthan, India. IV.Seventeen species as new record from the State. *Records of Zoological Survey of India*, 101(3 & 4): 129-145.

Bohra, P. and Baqri, Q.H. (2005). Two new species of Dorylaimida (Nematoda) from India. Indian Journal of Nematology, 2: 107-111.

Bohra, P. and Sultana, R. (2008). Four new species of nematodes (Nematoda: Dorylaianida and Isolaimida) from Rajasthan, India. *Records of Zoological Survey of India*, 108 (Part-1): 1-13.

Bohra, P. and Sultana, R. (2010). Eleven species of saprophagous nematodes as new record from Rajasthan. *Records of Zoological Survey of India*, II 0(4): 9-17.

Bohra, P. and Sultana, R. (2012). Nematodes of Thar Desert, Rajasthan. *Records of Zoological Survey of India*, Occ. Paper No., 334: 1-95.

Charan, P. D. and Singh, M. (2018). Invasive alien species of flora of the thar desert of Bikaner division, Rajasthan. *Journal of Emerging Technologies and Innovative Research*, 5(4): 568-572.

Chauhan, S. S. (2010). Mining, development and environment: a case study of Bijolia mining area in Rajasthan, India. *Journal of Human Ecology*, 31(1): 65-72.

Chhangani, A. K. (2017). First Record of the Pied Crow (Corvusalbus) from India, in and around Jodhpur, Rajasthan. *Indian Forester*, 143(10): 1081-1083.

Chhotani, O.B. and Ray, K.K. (1976). Fauna of Rajasthan, India. Hymenoptera. *Records of Zoological Survey of India*, 71 (1973): 13-49.

Choubisa, S. L. (2001). Endemic fluorosis in southern Rajasthan, India. *Fluoride*, 34(1): 61-70.

Choubisa, S. L. (2010a). Fluorosis in dromedary camels in Rajasthan, India. *Fluoride*, 43(3): 194-199.

Choubisa, S. L. (2010b). Osteo-dental fluorosis in domestic horses and donkeys in Rajasthan, India. *Fluoride*, 43(1): 5-12.

Datta, A. K. and Majumdar, N. (1970). Fauna of Rajasthan. India. Part 7. Fishes. *Records of Zoological Survey of India*, 62(1-2): 63-100.

Faisal, M. and Ahmad, M. (2005). Collembola diversity in changing agricultural landscapes of the Thar Desert: A case study in Sriganganagar district of Rajasthan, India. *Journal of Arid Environments*, 63(4): 717-724.

Gupta, P.D. (1973). Fauna of Rajasthan, India. Part 3. Mollusca (No. 2). Distribution of some Pelecypoda. *Records of Zoological Survey of India*, 67 (1969): 159-163.

Gupta, P.D. and Agrawal, V.C. (1966). Geographical distribution of Indian Hairy-footed Gerbil, Gerhillusgleadowi Murray. *Science and Culture,* Culcutta, 32: 470-471.

Gupta, P.D. and Agrawal, V.C. (1968). Cannibalism in Indian Five-striped Squirrel, Funambuluspennanti. *Science and Culture*, Calcutta. 34: 185.

Gupta, V. (2012). Mammalian scat as a Bio-Indicator of Heavy Metal contamination in Western Rajasthan, India. *International Journal of Scientific and Research Publications*, 2(12): 1-7.

Gupta, V. (2013). Mammalian Feces as Bio-Indicator of Heavy Metal contamination in Bikaner Zoological Garden, Rajasthan, India. *Research Journal of Animal, Veterinary and Fishery Science*, 1(5): 10-15.

Gupta, V. and Bakre, P. (2013). Heavy Metals Contamination in Mammalian Wildlife of Talchaper Blackbuck Sanctuary vsDhavadoli Protected Area of Western Rajasthan, India. *International Journal of Scientific and Technology Research*, 2(1): 86-91.

Idris, M. O. H. D., Singh, P. and Johari, S. (2009). Impact assessment of the Indira Gandhi Canal on the Avifauna of the Thar Desert. In: *Faunal ecology and conservation of the Great Indian Desert* (pp. 119-135). Springer, Berlin, Heidelberg.

Kankane, P.L. (2000). Status survey of Chinkara and Desert Cat in Rajasthan. *Records of Zoological Survey of India, Occ. Paper*. 179: 1-71.

Khana, V. (2005). Scolopendrid centipedes of desert national park, Rajasthan (Chilopoda: Scolopendromorpha). *Records of Zoological Survey of India*, 104: 121-127.

Kidwai, Z. and Bhattacharjee, S. (2016). An observation of Greater Flamingos (Phoenicopterusroseus) in Chicharli wetland, Dhawa, Rajasthan, India. *Zoo's Print*, 31(3): 1-2.

Krishna, A. K. and Govil P. K. (2004). Heavy metal contamination of soil around Pali Industrial Area, Rajasthan, India. *Environmental Geology*, 47:38-44.

Krishna, D. and Prakash, I. (1955). Hedgehogs of the desert of Rajasthan. Distribution and fossorial habits. *Journal of Bombay Natural History Society*, Bombay, 53: 38-43.

Krishna, D. and Prakash, I. (1956). Hedgehogs of the desert of Rajasthan. Food and feeding habits. *Journal of Bombay Natural History Society*, Bombay, 53: 366.

Krishna, D. and Prakash, I. (1960). Hedgehogs of the desert of Rajasthan. Pt. 3. Food in nature. *Proceedings of Rajasthan Academy of Science*, Pilani, 7: 60-62.

Krishna, D. and Prakash, I. (1960a). A note on mammals of Rajasthan desert [in Hindi], *Vijnana Parishad Anusandhan Patrika*, Allahabad, 3 (4): 203-214.

Kumar, A., Sharma, L. L. and Aery, N. C. (2008). Physicochemical characteristics and diatoms as indicators of trophic status of Kishore Sagar, Rajasthan. *Proceedings of Taal*, 1804-1809.

Mahajan, K.K. (1963). Fauna of Rajasthan, India: Protozoa No. 1. *Records of Zoological Survey of India*, 61(2):377-401.

Mahajan, K.K. (1971). Fauna of Rajasthan, India: Protozoa No.2. *Records of Zoological Survey of India*, 63 (10): 45-76.

Mahajan, K.K. (1977). Fauna of Rajasthan, India: Protozoa No. 3. *Records of Zoological Survey of India*, 72: 213-225.

Meena, D., Sharma, V., Kachhawa, B., Jangid, A. K., Senapati, T. and Yadav, D. (2018). Mining a Threat to Ajmer Avifaunal Biodiversity. *Journal of Civil Engineering and Environmental Technology*, 5(4):231-241.

Mohnot, S.M. (1977a). Observations on maternal behaviour in the langur, Presbytis entellus, in India, In: *The Natural Resources of Rajasthan.* (Ed. M.L. Roonwal), Vol. I, pp. 499-503. Jodhpur University Press, Jodhpur.

Mohnot, S.M. (1977b). Interactions and social changes in troops of the langurPresbytis entellus in India. In: *The Natural Resources of Rajasthan* (Ed. M.L. Roonwal), Vol. 1, pp. 505-514. Jodhpur University Press, Jodhpur.

Mohnot, S.M. (1981). On langurs of Jodhpur. Abstracts natnl. *Sympos. Evaluation of Our Environment* (Jodhpur, Apr. 1981), Jodhpur: 47.

Prakash, I. (1956). Additions to the list of mammals of Rajasthan Desert. *Journal of Bengal Natural History Society*, Darjeeling, 28: 169-170.

Prakash. I. (1958). Extinct and vanishing mammals from the, desert of Rajasthan and the problems of their preservation. *Indian Forester*, Dehra Dun, 84: 642-645.

Prakash, 1. (1962). *Ecology of the gerbils of the Rajasthan Desert*, India. Mammalia, Paris, 26: 311-331.

Prakash, I. (1963). Taxonomical and ecological account of the mammals of Rajasthan desert. *Annals of Arid Zone*, Jodhpur, 1: 142-162.

Prakash, I. (1977). Biology of the rodents of Rajasthan Desert. In: *The Natural Resources of Rajasthan.* (Ed. M.L. Roonwal), Vol. 1, pp. 337-352. Jodhpur University Press, Jodhpur.

Purohit, A., Barad, H. R. and Kumawat, R. (2020). Additional records of Greater Short-nosed Fruit Bat Cynopterus sphinx from the Thar Desert, India. *Zoo's Print*, 35(3): 4-7.

Rao, A., Poonia, S. and Choudhary, S. (2012). Climate change projections and impacts on Thar Desert ecosystem. *Ann. Pl. Soil Res*, 14(2): 87-94.

Rathore, N.S. (1986). New record of the termite Microcerotermesannandalei Silvestri (lsoptera: Termitidae: Amitermitinae) from Rajasthan. *J. Bombay Nat. Hist. Soc.*, 83 (1): 244.

Rathore, N.S. (1987a). New record of Cryptotermeshavilandi (Sjostedt) from Rajasthan, India (Isoptera: Kalotermitidae). *J. Bombay Nat. Hist. Soc.*, 84 (3): 694-695.

Rathore, N.S. (1987b). First Indian record of imago of the termite Eremotermesneo paradoxalis (Termitidae: Atnitertnitinae) from India. *Ann. Arid Zone*, 26 (3): 177-180.

Rathore, N.S. (1989). A new termite of the genus Microtermes (Isoptera: Macrotermitinae) from Rajasthan, India. *Entomon*, 14 (1): 53-57.

Rathore, N.S. (1994a.) Breeding biology of Indian Desert Termite Microcerotermes raja (Isoptera: Termitidae). *Ann. Arid Zone*, 33 (1): 61-66.

Rathore, N.S. (1994b). Observation on swanning behaviour of the termite Erelnoternlesparadoxalisin Jodhpur (India). *Entomon*, 19 (3 & 4) 135-137.

Ray, H.C. and Mukherjee, A. (1969). Fauna of Rajasthan, India. Pt. 3. Mollusca. *Records of Zoological Survey of India*, 61 [1963]: 403-436.

Roonwal, M.L. (1970). Termites of the Oriental Region. In: *Biology of Termites* (Ed. K. Krishna and F.M. Weesner), Vol. 2, pp. 315-391. Academic Press, New York and London.

Roonwal, M.L. (1972). *Termite Problems in India, Council of Scientific and Industrial Research*, New Delhi. Viii+81 pp.

Roonwal, M.L. (1973). Mound-structure, fungus combs and primary reproductives (king and queen) in the termite Odontotermesbrunneus (Termitidae) in India. *Proceedings of Indian National Science Academy*, New Delhi/ (B) 39: 63-76 (4 pls.).

Roonwal, M.L. (1975a). Thar Desert termites. In: *Environmental Analysis of the Thar Desert* (Ed. R. Guptri and I. Prakash), pp. 393-422. English Book Depot, Dehra Dun.

Roonwal, M.L. (1975b). Sex ratios and sexual dimorphism in termites. *Journal of Scientific and Industrial Research*, New Delhi, 34: 402-416.

Roonwal, M.L. (1975c). Field and other observations on the Harvester Termite, Anacanthoreter mesmacrocephalus (Rhodotermitidae), from the Indian Desert. *Zeitschriftfuerangewandte Entomologie*, Hamburg and Berlin, 78: 424-440.

Roonwal, M.L. (1977). Termite ecology in the Indian Desert eco-system: A review of recent work. In: *Desert Eco-system and its Improvement* (Ed. H.S. Mann), pp. 323-328. Monograph No. 1, Central Arid Zone Research Institute, Jodhpur.

Roonwal, M.L (1980). Termites and termite control in arid areas. *Lecture Notes ICAR Summer Institute on Management of Arid Zones in India*, Vol. 2, pp, 25-30. Central Arid Zone Research Institute, Jodhpur. (Cyclostyled).

Roonwal, M.L. and Bose, G. (1964). Termite fauna of Rajasthan, India. *Zoologica*, Stuttgard, 40 (No. 3, Heft 113), VI + 58 pp., 5 pls.

Roonwal, M.L. and Bose, G. (1969). Fauna of Rajasthan, India. Part 4. A checklist of Rajasthan termites (Insecta: Isoptera). *Records of Zoological Survey of India*, Delhi. 61 [1963]: 437-450.

Roonwal, M.L. and Rathore, N.S. (1974). Biological observations on three Indian Desert termites, Eremotermesparadoxalis, Microcerotermes raja and Angulitermesjodhpurensis (Termitidae). *Annals of Arid Zone*, Jodhpur, 13: 237-258.

Roonwal, M.L. and Rathore, N.S. (1975a). Swarming, egg-laying and brood-care in termites of genus Microtermes (Termitidae) in the Indian Desert. *Journal of Indian Academy of Wood Science*, Bangalore, 6: 37-55.

Roonwal, M.L. and Rathore, N.S. (1975b). Swarming, egg-laying and hatching in the Indian Desert Harvester Termite, Anacanthotermesmacro cephalus (Hodotermitidae). *Annals of Arid Zone*, Jodhpur, 14: 329-338.

Roonwal, M.L. and Sen-Sarma, P.K. (I960). Contributions to the Systemu-ticsof Oriental Termites, xiv + 407 pp., 65 pis. Indian Council of Agricultural Research, *Entomological Monograph* No. 1, New Delhi.

Roonwal, M.L. and Verma, S.C. (1977). Re-survey of the termite fauna of Rajasthan, India and its Zoogeography. *Records of Zoological Survey of India*, 72:425-480.

Sanyal, A. K. and S. K. De. (2005). Status of Ticks (Acari: Metastigmata) of Rajasthan. *Records of Zoological Survey of India*, 104 (Part 3-4): 129-136.

Sarang, N., & Sharma, L. (2004). Impact of Sewage discharge on water quality and benthic diversity of Kota Barrage, Kota, Rajasthan, India. *J Ind. Pollut. Contr.*, 30(2): 207-12.

Sewak, R. (2009a). Dung beetles (Coleoptera: Scarabaeidae: Coprinae) of Rajasthan. *Records of Zoological Survey of India*, 296: 1-106.

Sewak, R. (2009b). Dung beetles (Coleoptera: Scarabaeidae: Coprinae) of Rajasthan. *Occasional Paper-Records of the Zoological Survey of India*, (296), 1-106.

Sharma, G. 2015. *Pictorial Handbook on Damselflies and Dragonflies (Odonta: Insecta) of Rajasthan: 1-266, +93 photoplates*. Publ.: Director, ZSI, Kolkata.

Sharma, P. K. (2015). Wildlife Impact: Assessment and Conservation Measures on Schedule-1 Species to Development Projects like Mining and Cement plants, District: Chittorgarh (Rajasthan). *Energy Technology & Ecological Concerns: A Contemporary Approach*, 120.

Sharma, S., Sharma, S. K. and Sharma, S. (2003). Notes on mammalian fauna of Rajasthan. *Zoos' Print Journal*, 18(4): 1085-1088.

Sharma, V., Meena, D., Kumawat, R. K., Yadav, D., Kumawat, N. K. and Sharma, K. K. (2014). Mining and habitat encroachment: a major threat to the population of Hemidactylustriedrus at central Aravalli foot hills, Rajasthan, India. *International Journal of Environment & Animal Conservation*, 3(2): 1-8.

Singh, H., Marwal, R., Mishra, A., & Singh, K. V. (2013). Invasion of the predatory mosquito Culex (Lutzia) fuscanus in the western desert parts of Rajasthan, India. *Polish Journal of Entomology*, 82(1): 49-58.

Sinha, Y.P. (1973). Taxonomic studies on the Indian horseshoe bats of the genus *Rhinolophus Lacepede*. *Mammalia*, Paris, 37: 603-633.

Sinha, Y.P. (1975). New records of bats (Chiroptera) from Rajasthan. *Science and Culture*, Calcutta, 42: 608-610.

Sinha, Y.P. (1976). New record of Indian Sheath-tailed Bat, Taphozous longimanus, from Rajasthan, with remarks on winter fat deposition. *Science and Culture*, 42: 168-170.

Sinha, Y. P. (1980). The bats of Rajasthan: Taxonomy and zoogeography. *Records of Zoological Survey of India*, 76: 7-63.

Sivaperuman, C. and Baqri, Q. H. (2009). Avifaunal diversity in the IGNP canal area, Rajasthan, India. In: *Faunal Ecology and Conservation of the Great Indian Desert* (pp. 113-118). Springer, Berlin, Heidelberg.

Sivaperuman, C. and Rathore, N. S. (2004). A preliminary report on spiders in Desert National Park, Rajasthan, India. *Zoos' Print Journal*, 19(5): 1485-1486.

Sultana, F., Khan, S. and Sasmal, A. (2012). Assessment of water quality and pollution status of Chandralohiriver, Kota, Rajasthan. International *Journal of Pharmacology and Biological Sciences*, 6(2): 79.

Tak, N. (1995). Studies on Ants (Formicidae) of Rajasthan. *I. Jodhpur*, Hexapoda 7(1): 17-28.

Tak, N. (2008). Ants (Formicidae) of Rajasthan. *Records of Zoological Survey of India*, Occasional paper No. 288:1-54.

Tak, N. and Rathore, N.S. (1996). Ant (Formicidae) Fauna of Thar Desert: *Faunal Diversity in the Thar Desert: Gaps in Research*. Edited by Ghosh, Baqri and Prakash. Scientific publishers, pp. 271-279.

Tikader, B.K. (1966). On a collection of spiders (Araneae) from the desert areas of Rajasthan (India). Records of Indian Museum, Delhi, 59 [1961]: 435-444.

Vazirani, T.G. (1970). Fauna of Rajasthan, India. Part 5. Aquatic beetles (Insecta: Coleoptera: Dytiscidae). *Records of Zoological Survey of India*. Delhi, 62 [1964]: 29-49.

Verma, A. K. and Choudhary, A. (2017). Effect of heavy metal toxicity on Zooplankton population based on dyes and printing industries in Jodhpur (Rajasthan). *Biolife*. 5(1): 69-73.

Vyas, A., and Pancholi, A. (2009). Environmental degradation due to mining in South Rajasthan: a case study of Nimbahera, Chittorgarh

(India). *Journal of Environmental Research and Development*, 4(2): 405-412.

Yadav, V. K., & Chauhan, P. S. (2018). Avifaunal Diversity and Status of Jhalawar Forest Division, South-Eastern Rajasthan, India. *Indian Journal of Ecology*, 45(1): 107-116.

In: India
Editors: J. Lessard and O. Rivière
ISBN: 978-1-53618-413-6

Chapter 4

THE SOCIAL MASS PEDAGOGY OF TAMIL CINEMA: WITH ILLUSTRATION OF SELECTED FILMS OF THE MOST POPULAR AND PRODUCTIVE SCREENWRITER

Israel Selvanayagam
Professor Emeritus of Religious Studies,
Gurukul Lutheran College and Research Institute, Chennai, India

ABSTRACT

The Tamil people living in southern India and neighbouring regions, as well as in Diaspora (about 80 million), claim a unique pride in their Dravidian identity, their long history and the classic nature of their language. In post-Independence India (since 1947) they have had to struggle to safeguard their identity from many challenges. The Dravidian movement with its powerful proponents initiated a Tamil renaissance with their provoking speeches and writings, and cinema has played a major role. Consequently, the leading section of the movement captured political power in 1967, and one of their first activities was to change the name from Madras State (formerly Presidency) to Tamil Nadu (Tamil country). M. Karunanidhi (1924-2018), an avowed atheist and rationalist, popularly referred to as Kalaignar (artist), five times chief minister, an electrifying orator, with many honours including a doctorate, penned more than 70 film stories and dialogues (as well as 100 books), of which 39 hit the silver screen.

We have made a selection of unequal length depending on the focus of the story and length of the narrative in order to study their communicative power and to undertake a content analysis. A school dropout from an ordinary family, an activist from a young age, his sacrificial determination to fight against religious superstition and caste discrimination and for the protection of the pure Tamil language against the imposition of the northern Indian Hindi was phenomenal. The introduction of motion pictures (1940s) gave him a unique opportunity to propagate the ideas of justice and equality with reference to myths and folklore. Films proved to be one of the cheapest means of entertainment, education and conscientization of the masses, most of whom became supporters and members of his political party. Some of his film stories have been most effective to make people think and get organised with awareness of the need of political power for effecting change. However, in due course, unexpectedly, his road towards the peak became rough with two rivals who came to fame by the same medium of cinema as actors, with a similar overarching appeal to the masses. Perhaps his life ended with an anti-climax when there was a legal battle to secure a burial place in a prime spot along with that of his mentor and two opponents. Does it have to do with the pedagogic nuances of cinema or using it with a hidden agenda of dominant forces which employed the tested tactics of countering and confusing ideologies? Who influenced the people more, the playwrights behind the screens or the actors with strategic titles and roles on them?

After presenting a short introduction to the historical and cultural background we narrate in detail Karunanidhi's most decisive film *Parasakthi* (Supreme Power, 1952). It challenges certain cultural stereotypes (the sexual exploitation of women), social discrimination, corruption, religious superstition and so on. After an interlude of new challenges, *Neethikkuth Thandanai* (Punishment for Justice, 1987) narrates a fight against domination, injustice, corruption of the police and judiciary involving beastly rapes and gruesome revenges. *Puthiya Parasakthi* (New Supreme Power, 1996) corresponds to the film of 1952 but with greater violence. In between, a number of films are mentioned in passing which deal with certain aspects of social reality and their pedagogic power. The final part of the article, 'Highlights and Questions', has ten points of irony, frustration and depression.

A VIBRANT BACKGROUND

In the wake of India heading towards national independence (in 1947) after 300 hundred years of British rule, developments were not mono-focused and unilateral. With contributions from outstanding scholars and poets including some Christian missionaries there were ideological conflicts combined with the continued battle with the age-old caste system and Bramanic supremacy, along with religious superstition and grinding poverty and disabling illiteracy on a massive scale. Side by side with this, there was a search for ancient historical roots, regional identities and linguistic renaissance. No other region, cultural identity and authentic language came to gain prominence as southern India, Dravidian culture and Tamil language respectively. The full and integrated history of this movement has yet to be written, but for our limited purpose we will just point out the two mentors of the outstanding screen writer whose pedagogic approach is the subject of our study.

E.V. Ramasamy Naicker (1879-1973), a wealthy business man, a practicing humanist, a rationalist and atheist from Erode, a town in central Tamil Nadu, 398 KM from Madras (Chennai) came to be known as the Great One, *Periyar* (Diehl, 1978). He was for some time with the Indian National Congress which he later found to be dominated by a Brahmanic elite. M.K. Gandhi was not appealing as he was not clearly against the caste system. Periyar founded the Self-Respect Movement and worked for the abolition of caste and untouchability. He joined the Justice Party, but in 1944 transformed it into a mass movement with the new name *Dravida Kazhagam* (DK, the Dravidian Federation). He declared as the five pillars of his movement: 'No God, no religion, no Gandhi, no Congress and no Brahmin'. Even during his lifetime his statues were established with the following inscription of his declaration: 'He who created God is a fool, he who propagates God is a scoundrel, and he who worships God is a barbarian.' He went to jail many times for agitating against the imposition of the northern and Sanskrit-connected language Hindi, and for desecrating idols of gods and goddesses who in his view perpetuated superstition and untouchability. Through his newspapers and articles he established a new style of writing and speaking Tamil, purging it of the Sanskrit infection. However, he hated the medium of cinema as he thought it would lead to manipulation and the obstruction of rationalist thinking.

C.N. Annadurai (1909-1969), from a weaver community in Kanjevaram, a town 73 km west of Madras (Chennai), came to be affectionately addressed as 'Anna' (older brother). A brilliant student who achieved a rare triple MA and was awarded many prizes for essay competitions, Anna came to be called 'the Southern Gandhi', but was ideologically open within a liberation focus. He found a similar focus in ancient Tamil poems and in the writings of Tamil saints, poets and scholars (Kannan, 2010). Through his readings he discovered a Dravidian (southern) heritage that he saw as very different from the Aryan (northern) heritage which has tried to swallow the former. There is a scholarly credit given to him with an imaginative reconstruction of Tamil culture (Ryerson 1988, p. 105). Significantly, although an orator and prolific writer, Anna was positive about using the medium of cinema for conscientizing the Tamil masses, and he himself penned some stories which became famous. In any case, the Periyar-Anna duo did not work for more than 22 years, due to certain personal, ideological and methodical issues.

In 1949 Anna founded the DMK (the Dravidian Progressive Federation) as a vibrant party combining the core principles of the DK but with a view to getting political power in order to realize the Dravidian vision and ideals. He too went to jail for anti-Hindi agitation. His watchwords 'Duty, Dignity and Discipline' captured the imagination of the people. The challenge was to realize the ideals in a creative political system and just rule. The DMK swept through the elections and won in 1967. As had been promised, the name 'Madras State' was changed to 'Tamil Nadu', the poor got food rations at a reduced rate, the couplets of the famous *Thirukkural* of the ancient sage Thiruvalluvar were propagated as the verses of the Tamils' Scripture, and cleaning the massive sewage system of Madras was attempted. But, to the shock of his affectionate people and party colleagues, Anna died of cancer in 1969. It was no surprise that he attracted three million mourners for his funeral, as entered into the Guinness records. Perhaps no one was more shattered than his mentor Periyar and his affectionate brother M. Karunanidhi for whom he remained the 'deity of my heart' until the end of his life.

Muthuvel Karunanidhi (1924-2018), from the Isai Velalar community of Thirukkuvalai in Nagapattinam (former Thiruvaiyaru) District, 30 miles from Nagapattinam, dropped out of school in fascination for the Dravidian movement and he joined the Anti-Hindi agitation (1953), the climax of which, even defying the dictate of his affectionate mentor Anna, was his determination to have himself killed by a running train. Already influenced by the writings of many, he joined the DMK and won the affection of Periyar and Anna. Along with his many other achievements as mentioned above, he penned 70 film stories and dialogues of which 39 hit the silver screen. We have made a selection of the most popular in order to study their content and communicative power. His films too proved to be the cheapest means of entertaining, educating and possibly conscientizing the masses, most of whom became supporters and members of his political party. Some of his film stories have been so effective to make people think and get organised with awareness of the need of political power for effecting change. However, in due course, unexpectedly, his road towards the peak became rough as we will point out later.

Like many of his contemporaries, Karunanidhi started writing theatrical plays which became plays for films – mythical-historical-fictional. Already the 'theatrical drummers' played a major role in creating socio-political-national awareness directly and indirectly, often with satirical sarcasm (Baskaran, 2020). Cinema, however, came to provide wholesome entertainment with music. Also, while the British were projected as the main enemy, particularly by robbing the natives in the form of tax, creative writers came to realise evils like caste, poverty and the oppression of women remaining. Karunanidhi's following film best illustrates the wholesomeness of entertainment, education and enlightenment.

PARASAKTHI (SUPREME POWER, MOTHER GODDESS, MADURAI MEENAKSHI)

Though the story was from a theatrical play, Karunanidhi's fifth film *Parasakthi* (1952) proved to be decisive and most influential. Its sharp challenge for religion, society and the government of his day meant that there was initial opposition which demanded re-censoring. When it was declared passed there was euphoria and even mass hysteria to view it again and again travelling long and wide, often by foot.

The Core Story

The background of *Parasakthi* was the devastating effects of the Second World War in 1942. Three brothers of a family (Chandrasekaran, Gnanasekaran and Gunasekaran) were in Rangoon (Burma) as immigrant workers. Their father and his only daughter Kalyani were living in the southern temple town Madurai of the presiding deity Meenakshi or *Parasakthi*, in constant anxiety about the life and safety of the three men. Socio-cultural expectations forced the father to arrange a marriage for his daughter, incurring huge debt. At the time of Kalyani's first delivery her husband lost his life in a road accident; soon after, the father also passed away. Having lost her property to the lenders, after a traumatic moment, Kalyani looks after her baby boy while running a street corner snack shop for their survival (as was the chore of widows). The difficulties she faced as a young widow includes some men asking her for sex.

The brothers in Rangoon knew about their beloved sister's wedding, but they could not all go because of seat limitations on the passenger ship, so they decided to send the youngest Gunasekaran who travelled with gifts for his sister. On his arrival in Madras (Chennai) he lost everything to a prostitute who he met in a hotel, and he became a pauper on the street where he experienced the cruelties of the society experienced by the penniless. Finally, he pretended to be mentally insane in order to justify his acts such as stealing for survival, and with the same condition finally took a train and reached Madurai. His meeting with his sister in her pitiable condition without her knowing his real identity breaks his heart. He follows her discreetly and saves her from predators.

In the meantime, the two brothers, one with his wife, joined the Indian refugees returning home because of the frightening and shattering war, and in due course they too were separated from each other. Chandrasekaran became a judge in the court in Madurai. Gnanasekaran, disabled on crutches, after failing to get a place in the Tamil refugee camp, became a beggar, but an enlightened beggar who challenged the ruthless officer of refugee camp for his inappropriate behavior, and promoted cooperation amongst the beggars, spearheading a human rights campaign for them.

Kalyani went through several ordeals. She was employed by a general merchant who showed all malpractices such as hoarding, but was a devout Hindu. Plotting to send his wife and a male servant away, he tried to rape Kalyani, but she escaped and he was caught red-handed by his instantly returning wife. As a beggar carrying a child in starvation Kalyani unknowingly approached Chandrasekaran's house while a party was going on. She was rebuked and when she embraced his feet for mercy she was kicked out and he (her brother) justified himself for not showing mercy, saying he was not able to see his own sister. Kalyani checked in the refugee camp if her brothers were there and was directed to go to the beggars' group nearby. When she went there Gnanasekaran was out for a finance campaign, and she enquired with the group and they had no clear answer. When he returned to the group they told him about a young lady with a child asking about her three brothers. When pressed for their names they told him the names of all the three brothers. Gnanasekaran also started his search for her. In a way now Kalyani had met all of her brothers without knowing them, but the encounters reveal how they were unfortunate in not being able to see their sister, and for that very reason their treatment of her also was crude. The case of Gunasekaran was different. He in disguise of a madman follows her and rescues her from those who wanted to use her for sex.

Finally, Kalyani takes refuge in a temple where the priest tries to molest her. The assistant priest, the bell-ringer, who has previously met Gunasekaran, helps her to escape by ringing the bell vigorously. Through him Gunasekaran comes to know about the priest's misbehavior with his sister, and he appears inside the temple during a *puja*, creates a disturbance and attacks the priest, which leads to his arrest and prosecution in court.

Having tried every resort and failed, Kalyani decides to throw her child into a river and to commit suicide by herself jumping into the water. The baby is thrown but just before she jumps she is arrested by police and taken to court with charges of murdering a child and attempting to commit suicide. She is presented before judge Chandrasekaran. While Kalyani is narrating her story and justifying her action in the given circumstances, Chandrasekaran gradually realizes with surprise and astonishing marks on his face that she is his sister. The climax is judge Chandrasekaran falling from his chair and passing out. He loses his memory and recovers at home with medication.

In the meantime, Gunasekaran is encountered by Vimala, a well-off woman, enlightened by her older brother, a great thinker. Gunasekaran remembers the previous act of once stealing her food and running away and thinks that he is now caught, but later realizes that she is of a different type. She picks up the conversation referring to his song meaning 'Don't trust even the rationalist world'. She takes him home and listens to his whole story. While sympathizing with his ordeals, she challenges him to think of masses in the society who go through such ordeals every day. She exhorts him to see the root cause of the problem instead of seeing its outward manifestation. In the course of his stay in her house overnight and further conversation she develops some interest in him, but he leaves her without giving his full consent.

Another judge had led the case of Kalyani, but now Gunasekaran joined the trial, and at first they did not acknowledge each other as brother and sister. Gunasekaran answered and argued on behalf of his sister but when the judge intervened saying that it was the case of another person, Gunasekaran revealed that Kalyani was his sister. But Kalyani refused to accept this, and said that her meeting him in the past was only as a 'mad-brother'. When her act of 'murdering her child' was mentioned Vimala appeared with the child saying that she had got the child when he hit her boat during a moon-light boat ride and was bringing him up in her home, and when her brother on his return from travel abroad pointed out about the ongoing trial of Kalyani from the newspaper she had rushed to the court. She also confirmed that she knew the whole story and that before the court standing on trial were Gunasekaran and his sister Kalyani. Emotions ran high and Gunasekaran justified all his acts in the temple.

They were acquitted as exceptional cases in a distinctive situation with reference to similar acquittal in other courts. Finally, the family was reunited at Chandrasekaran's place in highly-charged waves of emotion. He recovered fully. At last, Gnanasekaran appears on his crutches in front of the house of Chandrasekaran where all the rest of the family and Vimala were, and asked for a donation for the proposed beggars' conference. Through knowing the watchman from their refugee days, he also re-joined the family, and joy overflowed. Finally, Gnanasekaran was wanting to go back to his wandering life, but was persuaded by Vimala to make the beggars bungalow-dwellers, not vice versa. Also, her dream of becoming the wife of Gunasekaran was now realized. A beggars' conference was organized. The chief-guests included Periyar and Anna. The film ends with an affirming chorus meaning 'We all should live according to the vision of our good forebears.'

The Sharp, Insightful and Challenging Statements and Questions

Karunanidhi proved himself a genius by screen-writing a core theatrical story with succinct and thought-provoking dialogue. The way he portrayed society and its issues had some appeal to every person who watched the film. We will illustrate this by reproducing the most significant parts of the dialogue in the film which mention the most fundamental aspects of human life in Tamil society.

Poverty

The terrible nature of poverty is represented by the increasing number of beggars and the magnitude of their problems. On his arrival in Madras, the first encounter for Gunasekaran was with a beggar asking for alms, which disgusted him. His own experience as a beggar which pushed him to be a 'mad-beggar' taught him lessons about the suffering of beggars and the poor people in society. He later declared poverty a ghost, and there was no way to cast it out. He also observed that pets like dogs were well fed, but not a widow like Kalyani and her starving child.

Gnanasekaran, as one of the disabled refugees, created solidarity among the beggars and mobilized their voices for recognition and remedial measures. Hence the conclusion of the film with a beggars' conference to which top figures were invited. Was this a solution?

Kalyani wandered around with her child and begged, but not only were her feelings hurt by the indifferent or cruel responses but also her spirit was weakened and her psyche wounded. She said in the court: 'My child was suffering from hunger. Let me say this very frankly for I consider those around me are my brothers and sisters. Until his mouth went dry my son wetted my blouse. Having gone without food for months how could I have milk for him? I could not bear the torture, so I killed, is it a crime?'

Kalyani encountered a family living in poverty and starvation but who were not willing to come to the street. Certain self-dignity, and perhaps the ownership of their house, kept them inside. The head of the family guided her to go to the rich family where a party was going on in the same street.

Society

Society is made up of people, cultures, customs, values, relationships, attitudes and remedial structures. The victims in *Parasakthi* experienced society in a particular way. Vimala was right when she challenged Gunasekaran that if he had not been pushed to the limit he and his peers would not have known the dynamics, evils and ambivalences of their world. She points out the fact that Gunasekaran's love and kindness were not matched by analytical knowledge, for he should have first gone to the police when he was robbed in Chennai. Society can be sometimes a 'hostel of mad people' and a 'cave of thieves', and he had also become a fool by joining this kind of society.

Vimala's further points of exhortation to Gunasekaran were more provocative: 'You are selfish to think of only your sister: how many such sisters have to struggle in this inhuman society! I do appreciate that prostitute who robbed you because she only opened your eyes. You try to control a volcano by pouring water! You play a tune to tame a giant serpent! Instead of cutting the branches of an unwanted widespread banyan tree we should uproot it with all its bonds and connections. No point in playing the fiddle like Nero when Rome is burning! Not to be idealist, let us be realist and work for a new society.'

Vimala's brother in his public speech mentions: 'It seems the world beauty Cleopatra drank wine with Tamil Nadu pearls! But Tamil women are dying of hunger, waiting for money from husbands working abroad! Brothers, we hear that some mothers are selling their children in hunger pangs! Their dreams of their children's childhood enjoyment, school, employment and marriage – everything is shattered! Now we have the case of Kalyani before us …'

Kalyani in court challenged the society, which tried to exploit the vulnerable condition of an individual. 'I rolled in poverty. My youth was a hurdle to seek a means for survival! I didn't fall into the traps of heartless crooks. Had I agreed I would have been living in bungalows, not as a sorrowful widow, but as a prostitute! That's what society wants, isn't it? That's how society can be tackled, is it? My time in particular social circumstances has made me an orphan!'

Before telling his real story in the court Gunasekaran says: 'Why are you so concerned when no one else in this world is concerned? You will ask like that. I was affected, I was personally affected. You may accuse me of selfishness. Public welfare also is entwined in my selfishness. This is like a fish that by eating the dirt cleans up the lake …'

'It's like that. They are calling me criminal. If you go through the past life of this criminal, you'll know the number of wild streams he had to cross in his life. There were no singing nightingales in my life's path, it was filled with snakes with their hoods ready to strike, I never enjoyed this cool breeze, I have crossed raging fires; listen to my story, before writing down the verdict – please listen.'

Then Gunasekaran summarises the suffering of his sister Kalyani and points out how society treated her: 'I saw my sister like a painting without eyes. Yes, as a widow. My sister's name is Kalyani, an auspicious one, yet there is no auspicious thread around her neck. The family that lived in riches got devastated. With a child in her hand and tears in her eyes, Kalyani roamed, and I roamed for Kalyani: many showed her favor, but some rogues among them demanded her love as a price. This scoundrel Velu, who's standing before you accused of murder, he tried to trap my sister with pomp and show. Had I not stopped him, my sister would've committed suicide then itself; she didn't want to die leaving her child in this heartless world. She didn't want her child to die helpless in agony, and therefore she herself killed it. To kill loved ones is not anything peculiar. Mahatma Gandhi, the embodiment of non-violence, the exponent of compassion, had himself ordered to kill a calf suffering from painful disease, seeing the suffering it was undergoing. Kalyani had also followed the same way. How can this be a crime? There is no way to live here for a young man returning from abroad, there's no protection of life for a girl born in Tamil Nadu. Had my sister only yielded a little, one day in a millionaire's bedroom, another day in the lap of a wealthy man who put a price on her hunger, she could have passed her days like this. Is this what this court desires from her? Pomp pursued my sister, she ran in fear; money chased my sister, she ran away again; religiosity threatened my sister, she ran … she ran to the brink of her death. She should have been stopped, her fears should have been removed by the people who quote Law now. Did they do it? Did they allow my Kalyani to live?'

When the judge points out 'The accused is becoming an advocate for somebody else's case,' Gunasekaran continues, 'No, not somebody else's case, it is also my case. It's my sister's case. What is wrong in a brother rushing to teach a lesson to a rogue who tried to molest my sister? Kalyani's attempt to commit suicide is a crime. Kalyani's killing the child is a crime. It's my crime to attack the priest. Who is the cause behind all these crimes? Who? Who? Who is the cause? Whose crime was it that allowed Kalyani to roam without food? Was it the fault of fate? Or was it the fault of worthless people who thrive in the name of fate? Whose fault was it that allowed a group of swindlers to grow ... was it the fault of famine? Or is it the fault of those who invited famine to their bed? Whose fault is it to make amorous fake priests become members of society? Was it the fault of God or that of the deceivers who thrive in the name of God? There'll be no dearth of myself and Kalyanis till these crimes are removed. This is the lesson, the wisdom and useful political philosophy from our life history.'

Religion

The religion involved in the story is Brahmanic Hinduism, which has influenced the regional devotional traditions and popular cults as well. The title itself is carefully chosen as *Parasakthi* which literally means 'Supreme Power', and normally it is used for the Goddess in her multifarious manifestations including Madurai's Meenakshi. However, the challenge here is applicable to all religions, which have direct influence on society.

The fruit-vendor from whom Gunasekaran stole a piece of fruit to satisfy his burning hunger, lamented saying, 'Let God hear you!' Gunasekaran replied, 'He has not heard my cry of dereliction! His heart has become stony.' In his disguise as a madman, Gunasekaran in his ecstatic mood sings the famous song of the nationalist poet Subramanya Bharatiyar:

> Regarding a stone-post as a deity and offering flower
> circumambulating and muttering mantras, what is it?

> Can the stone-post speak while the Lord is within us,
> Can the stirring spoon taste the curry of the pan?

The afore said merchant is portrayed as an example of nominal piety while his servant appears good-hearted. While they come down from a temple situated on the top of a hill, a beggar encounters them at the bottom and asks for alms. The servant gives him part of a broken coconut, but the master beats the servant for daring to give the 'sacred food' to a beggar. The servant reminds him of what was preached in the temple, i.e., 'A righteous act will protect one's head', and the master's angry response was 'Was that for you and me, fool?' Subsequently, another beggar forcibly took and ate some of the sacred food for which he was beaten up very badly. Gunasekaran appears and punishes the merchant and exposes his hypocrisy.

Gnanasekaran in his speech to the beggars tells them that there is no use in calling upon God in his abode: their voice must be raised to the government. And Vimala's brother, while narrating the pitiable condition of the poor in the society, mentions:

> Temples are here, the god has a diamond crown, golden armour and a ruby throne. Children of these gods, however, are struggling for a square meal.

The challenge to religion reaches the climax in the temple when the priest tries to molest the desperate Kalyani, leading her to her final decision to kill her boy and kill herself. Initially the priest welcomes her and encourages her to tell the Goddess all her problems. Then he slowly approaches and touches her, and when Kalyani protests, he trivializes the act saying that just like doctors have the freedom to touch their female patients, priests also have the freedom to touch devout women. She points out that such an act is unbecoming of a priest and his cynical response is, 'Is it that why all such things happen?'

During a festival time, when a crowd comes for prayer and acts of vow-fulfilment, and while the priest is reciting the mantras in the particular case of a victim of chicken pox, Gunasekaran appears in the sanctuary behind the idol of the Goddess. Following is the dialogue between the two:

> G: Hey priest, first check your own horoscope!
>
> P: Who, is it Mother speaking?'
>
> G: When did the Mother speak, you shameless one?
>
> P: Goddess, Parasakthi. Mother, what are you saying?
>
> G: The Goddess is your Mother, and my sister Kalyani is your prostitute, are you not one without any dignity?
>
> P: Who are you, and what is wrong?
>
> G: You had played with my sister's life without expecting any trouble. As the one who blesses devotees, have proved yourself a scoundrel! Now you stand in the box of the accused!
>
> P: Parasakthi, Parasakthi!
>
> G: She won't speak. If she could talk, when you had attempted to molest my sister, wouldn't she have stopped you by yelling at you, you senseless brute? My sister was caught up in your forced embrace, it seems she had called to the Goddess Paraasakti a thousand times. Did she come and offer protection to her?
>
> P: Oh devotees, why are you simply watching this?
>
> G: Why do you seek human support while you are a devotee of the Goddess, and while she is holding a trident, a sword and a mace?
>
> P: Transgression, violation!
>
> G: Hey priest, I dare death. I know what would be the reward for my deed: gallows, gallows! Yet let my courage be a lesson for this decaying society who thinks priests like you are pious men!
>
> P: Hey, why are you creating trouble in the festival of the Goddess? (*takes up his sword and tries to attack G but the latter plucks it from him*).
>
> G: Festival, festival, you and your goddess have ruined my sister's life; do you need a festival? (*attacks the P with the sword*). Behold, it is the festival for this knife, festival of bravery, and festival of victory!

Having escaped from the temple, Kalyani comes to her final act of throwing the child into the river and jumping in herself. The final dialogue with her baby is most poignant drawing upon popular notions about life after death. She tells her child: 'My beloved baby, this world is not going to give you life and therefore you go to the other world where there is no poverty. You appear to ask me, "Why do you take my life with the excuse of a life in the other world!" But I can't let you live alone in this fraudulent world'.

In the course of her justification for 'killing' the child Kalyani uses mythological stories: 'That Thirutthondar (a Saiva saint) who cooked his son for food is talked about with pride; why was it not prohibited? What (such saints) taught I have done and then how can this be a crime?' Further, when the judge insisted that she should have tried to save the child whatever the difficulty in life was, she responds, 'Is my child Thirugnana Sambandhar (another Saiva saint) to be fed by Goddess Parvathi? Only myself had to take care of him.'

In the last court scene, Gunasekaran mentions in his argument: 'The court has faced many peculiar cases. And has met many peculiar personalities. But this case is not peculiar. I'm also not a peculiar man who's pleading this case. I'm an ordinary man, whom you meet on street in everyday life. I created trouble at the temple, attacked the priest. I've been charged like this. You will be expecting me to deny all these. I won't, certainly I won't; I did create trouble at the temple, not because there shouldn't be a temple at all, but because it shouldn't become a den of crooks. I did attack the priest, not because he is a devotee, but to condemn devotion becoming a guise to cheat … even devotees of God came forward to protect Kalyani, demanding her personal devotion to themselves. The prime among them is this priest. He demanded her honor as a price in the name of Goddess Parasakthi, in the name of Mother of the World. Kalyani would have decided to be alive at least as a suffering worm in this world, but it is this priest who provoked her to commit suicide'.

Government

India had been declared a Sovereign Democratic Republic with a resolution to secure to all its citizens: Justice, social, economic and political; Liberty of thought, expression, belief, faith and worship; Equality of status and of opportunity; and to promote among them all Fraternity assuring the dignity of the Individual and the unity of the Nation'. This came into operation with the constitutional guidance on 26 January 1950, two years before Parasakthi was released. It was still new for the people of India to understand and abide by the principles on which the excellent Indian Constitution was built.

In the story the question of government came when Kalyani was arrested by the police and presented to the court. In the context of claiming that Kalyani's attempt to kill her child was a crime, the judge claimed that everything that is created belongs to the government. Kalyani's response is very succinct: 'Belonging! We struggled to get a mouthful of rice, we wandered and were rejected. We were treated worse than a sickened dog! We wriggled and withered like worms of starvation. During all those times why didn't the government help with its "belonging"? After deciding not to live amidst injustice and to get freed in a second if one goes to die, stretching its sceptre-like hands and telling stories of belonging, the government stops that decision! Wonderful government and marvellous justice!'

The Justice System

When in the court the judge claimed that Kalyani's 'killing' of the child was a crime, she replied, 'If that is the case, what about the story of Nallathangal who threw all her seven children into a well? That folk-story is played all over the country, and then why there is no law to ban it?' But the judge rebukes her saying, 'Don't quote such stories as precedent. Your anger should not insult justice, other than the facts of the case. Accusing the government or the society here is impertinent!'

Later the judge said that Gunasekaran's argument seemed unconnected. 'No individual has any right to punish a criminal. There are laws to take care of it. The right is given to you to defend your case … you mustn't use it to criticize others or society. Answer only the questions that are put to you. Is it true that Kalyani is your sister?

G: Doesn't this question sound funny even to you?

J: Alright, let Kalyani be presented here. (She arrives; *to Kalyani:)* Is this your brother?

K: No, I call him mad brother. He is not my own brother.

G: Kalyani, I am your own brother. I am Gunasekaran.

J: What is it?

K: He is slightly out of his mind!

G: Yes, I acted like a madman to avoid burdening my sister with further misery. Your honor, you need not believe me. You write out any judgment for me. But before that, a request. Please forgive my sister Kalyani.

J: You shouldn't talk about it. It is a crime to kill a child. (At this point the child is brought by Vimala, the confusion is cleared, and the family is reunited).

CONTINUED IMPACT

Apart from being viewed again and again in cinemas, the dialogue of *Parasakthi* has been broadcast everywhere. Several people could repeat it by heart and entertain the illiterate masses. Though there was no empirical study of the impact on individuals undertaken, no one would deny that the film's contribution to the fast growth of the newly established DMK (1949) and its capture of state power for the first time in 1967 was phenomenal.

Given the context of the post-Independence Indian society, particularly the situation in Tamil Nadu, *Parasakthi* presents a wholesome picture, raising questions about various aspects of life in society and leading to a solution. Though one can be emotionally carried away by the tragic situation of a family which reaches a happy end, for those who can identify with the story and look at the society with a critical mind, no doubt this film is a pedagogic tool to conscientize the poor and oppressed. It evaluates society, affirming the positive and denouncing the negative with a special focus on poverty, viewing critically the role of religion, government and the judiciary.

Karunanidhi refused to be categorised as a playwright of any particular literary genre. As we have noted, his stories were theatrical, historical, mythological and fictional. In every story, there will be references to the ancient Tamil literature and common human values such as justice, equality, dignity, sibling affection, parental affection, confusion-clarity, separation-union, crookedness-straightforwardness and so on. At the same time, *Parasakthi* may be taken as providing a framework and template using which we are able to elaborate and build up the issues with illustrations from some other film stories and dialogues of Karunanidhi.

EXAMPLES OF SOME RELATED THEMES

In the following list of films, the theme of *Parasakthi* is complemented with significant notes of twists, new nuances and even confusions in the pedagogical process of Tamil cinema.

Panam ('Money', 1952), depicts a greedy man who chases away the poor bride chosen by his son since she did not get the dowry that he demanded. While the girl is trapped in a deserted bungalow, her husband is married off to a rich girl.

Thirumpip Par ('Look Back', 1953) was one of the earliest Tamil films whose dialogues were known to be a political satire on the Indian National Congress, then the ruling party.

Poompuhar (based on one of the Tamil epics, 1964) focuses on the false accusation and murder of a jeweller by a king in Madurai; his wife Kannaki vindicates her husband and revenges the king and burns Madurai down. Kannaki came to be celebrated as the goddess of justice.

Poomalai ('Garland of Flowers', 1965) narrates the title character, a happy-go-lucky girl whose life changes for the worse when she is raped. Karunanidhi uses the story ingeniously, weaving it with ambiguous contemporary political processes and social evils into mesmerising dialogues with poetic punches. Sibling affection with mutual dedication is repeated. One memorable scene is a dialogue where the sister speaks highly of her brother and threatens to shoot anyone speaking ill of him.

AN INTERLUDE: CHALLENGES AND COMPROMISES

If Tamil cinema played a major role in social pedagogy with the decisive contribution of Karunanidhi, the process and methods were not consistent. Perhaps, quite unexpectedly, in the heyday of the nexus of cinema and political rule there was a shattering development. M.G. Ramachandran (1917-1987), a high caste Keralite, who after living in Sri Lanka, came to Madras, and first joined the Indian National Congress. After a period of theatrical acting, he joined the cinema and interestingly acted in the three films of Karunanidhi before 1952. With his help, MGR joined the DMK in 1953 and won a legislative seat, and was made Treasurer of the party in 1969. Initially MGR struggled to find a place in Tamil society, and his adoption of the Tamils' deity Murugan as his personal one and titles such as 'I am the child of your family' were not insignificant. But very soon he became a rising star not only as a matinee idol but also a charismatic 'revolutionary' leader (Kannan, 2017). It may be a worthwhile study to find out how much his projection on the silver screen as a promising leader and friend of the poor in different segments of the society contributed to his making. The premature death of Anna in 1969 left a vacuum for a commanding leader of reconciling factions and forces.

MGR split from his long-term friend Karunanidhi who used to address him 'my lord' and formed a new party which still kept the title 'DMK' with the added prefix of 'All India Anna' (i.e., AIADMK). In 1977 MGR became the first cine actor in India to become a chief minister. The DMK and the AIADMK continued to be contestants for political power in Tamil Nadu. After MGR his cinematic pair and political heir, J. Jayalalitha (1948-2016), a Brahmin from the neighbouring state of Karnataka, entered the Indian cinema with the distinction of being the first to appear on the silver screen wearing skirts. Initially it was the celluloid attraction that won her the admiration of masses. Following her first acting with MGR in 1965 she not only grew in fame as an attractive actor but also took (or was given!) the same title 'revolutionary leader'. The same way she inherited the title *Amma* (Mother) and this title continues to be prominent (Vaasanthi, 2016). On the whole, she came to fame by acting and connecting with MGR, whereas Karunanidhi was an established patriarch of the Dravidian heritage. Ordinary people may not know the difference between these two figures. Projected was a protracted political conflict between the two alternative chief ministers and their parties which led to nasty moments of political vendetta (web. 2020).

In the meantime, in the contest of film and fame, it seems that an implicit but significant difference arose. Karunanidhi was a playwright and MGR and Jayalalitha were actors. The screenwriter was behind the screen and he had no more prominence than having his name passing in a flash in the credits although Karunanidhi was popular as a speaker, writer, leader of the DMK and mature politician. But film viewers see the actors, and if their characters are in good roles they identify with them. Probably it was this situation that gave Karunanidhi the tactical thought of finding a space to introduce himself at the opening of his film. Already in *Poompuhar* (1965) he gave a succinct historical introduction in chaste Tamil but without saying anything about himself. The film *Kaalam Pathil Sollum* ('Time will Answer', 1980) reflected his satirical optimism for the future, but there was no speech. From *Palaivana Rojakkal* (1986) until the end in almost all his films he had an impassioned opening speech, narrating his affectionate attachment to the Dravidian movement with the double mentorship of Periyar and Anna, and his contribution to the literary, political and film world, as well as his anguish for the disorderly and corrupt nature of the Tamil Nadu government, particularly the police. There is no reason to doubt that these speeches helped him to keep his feet on the ground and connect with the people. Certainly, they helped him to arrest the growing political tide of MGR and Jayalalitha, and he could not be wiped out from political influence and power until the end of his life. Now we continue the selected films analysis.

Two Films of Intense Struggle

Palavaivana Rojakkal ('Desert Roses', 1986), a remake of the Malayalam language film *Vartha*, is about the freedom of the press, and the struggle between corrupt politicians and angry heroes. It focuses on a journalist, his lover and his friend as they fight a corrupt system to expose unscrupulous politicians. It is noticed that he used a far more subdued and sophisticated style to convey his political message but without compromising on the core of *Parasakthi*. Very interestingly, the godman placed on the side of the corrupt politicians is kept silent. Also, not simply struggling individually against evil, the playwright shows that people in different walks of life, including journalists and thinking individuals, can work together in a gentle yet revolutionary way.

Neethikkuth Thandanai ('Punishment for Justice', 1987) is arguably, next to *Parasakthi*, the most challenging film. A sensation was created by the rumor that the then government was going to ban the film, and that created a great rush to view it at the earliest opportunity. The box office proved not only the power of sensation and rumor but also an appeal to frustrated people with a unique blend of personal emotions, the corruption of politics and judiciary and the introduction of novel methods to bring out justice, sexual obscenity, and punishment for justice in a crudest form. The songs of a street singer both summarise what has been happening and provoke people as to what could be happening. He appeals to people's conscience, asking if it is dead, and asks the public why their hearts do not burst like volcanoes. If there is blood which does not boil, for him it is not blood at all. Finally, after successive defeats, the collective witness of the people is galvanized and truth is established amidst a bloodbath. And it was not surprising that the film was remade in other languages: in Telugu as *Nyayaniki Siksha*, in Hindi as *Kudrat Ka Kanoon* and in Kannada as *Nyayakke Sikshe*.

In a feudal context, the village chief Dharmaraja ('righteous king') is ruthlessly wicked and exploitative. He claimed that what the people think he speaks and what he tells the people should think! His influence helped to elevate his son-in-law Markandeyan to be a government minister, and their mutual support benefitted each. The chieftain's tyranny was challenged by an enlightened youth from the village who intervened in the unjust beating of a washer-man by his henchmen who wounded the youth severely. The only resort was to get a medical certificate of the incident. A newly arriving doctor to the village with his wife Chitra and their little girl, on request, agreed to give a certificate despite a warning. In a sudden ingenious conspiracy, the body of the youth is found in a cupboard of the doctor's clinic. In the court the washer-man is prepared to come up with a false statement. Accordingly, there had been an affair between the youth and Chitra and, being unable to bear this, the doctor killed him. In a torrent and whirlpool of emotions, the doctor is sent to jail and Chitra and her little girl are sent out of the village as she as a 'prostitute' would corrupt the society.

While Chitra with her girl went to complain in a police station she was raped by the police officer Chandrakandh, and when the child intervened he killed her by strangling her, which the mother reacted to violently. Then, carrying the dead body of her child, she goes to minister Markendayan. The minister expresses his shame that such things could happen in the rule of his party, promises justice and arranges the child's burial. But the next development was unexpected. With security guards manning the gates, the minister takes Chitra to his bedroom and violently rapes her. There is a punch line in the dialogue between the two security guards: 'Do you know who we are safeguarding?' and the other answers by shedding tears.

Escaping from the minister, Chitra went out with a plan to take her life by throwing herself under a passing train. In the nick of time she was rescued by a passer-by, a woman lawyer Thilkavathi known for her forthrightness, who took Chitra into her home. Her husband is an MLA (Member of the Legislative Assembly) and corrupt; once his car had hit a street singer and he escaped, and later Thilakavathi had to argue his case against her husband who was fined; this further strained their relationship and as well as his relationship with the minister. The singer also revealed that the same man had raped his wife which led her to commit suicide. At the same time, a newspaper reporter interviewed Markandeyan about the news of Chitra, and in the course of argument he was badly beaten up. Markendeyan justified his action by referring to Jesus' judgement of a woman caught in adultery: 'Let any one of you who is without sin be the first to throw a stone at her.' In any case, by twisting the facts, Markendeyan escapes the law and Chitra is punished with seven years jail.

A new plot and different twists: a foreign model with the name Sinduja arrives by flight and catches the lustful eye of Markendeyan. She has a strong bodyguard, and P.A. Markendeyan visits the 'queen' in her 'palace' and proposes marriage. Sinduja (*Chitra*) puts forward her conditions, and it happened that the personnel security guard was the police officer who had earlier raped her and killed her girl. Chandrakandh comes to sense the plot but Markendeyan finishes him off. Then he forcibly divorces his wife, which infuriates his father-in-law Dharmaraja, and he too is murdered. The CID officers come to arrest Markendeyan for two murders, but the court observes that there is no evidence; however, the lawyer Thilakavathy and others are ready to show the plots of secret burials (apart from confirmation of the girl's killing and burial, as even that had been challenged). Chitra's husband doctor is released, and his first encounter with his wife is in front of the minister. Shocked with usual suspicion, he heaps insult and guilt on Chitra, but when she tells him the true story of her disguise he is convinced. Now he too joins an ingenious plan for establishing justice. Further, a video taken by Sinduja's (Chitra) P.A. becomes strong evidence of Markendeyan's final struggle to rape Chitra.

The judgement is postponed to the next day. In the meantime, the judge is revealed to be the father of the senior lawyer Thilakavathy, and as she knows him to be corrupt which estranged her from him, she offers him a bribe to act justly in the case; however, he appears to be a changed man. The next day the newspapers announce that he has resigned. Soon her MLA husband returns to her announcing that he has been made a minister, and as he has already bought her father, now she has to withdraw from the case and go on holiday. When she refuses, he strangles her and kills her.

A new judge takes charge the next day. It is alluded that the video as the only evidence is unreliable as there is no way to check the actual face of the minister. There are dramatic reactions, as the blindfold of the goddess of justice in the court gets loosened and the scale on her hand loses its balance. Momentum gathers, a group of victims and their supporters are chasing the corrupt victimisers, and the final scene is Chitra piercing a sword through the belly of Markendeyan with intermittent flashbacks to his earlier cruel act of raping her. Finally, Chitra with her outstretched hands makes a solemn statement: 'If it is right that law has not punished these enemies of society, what we have done now is wrong, you should arrest us.' This is repeated by the people around.

FOUR FILMS IN NON-CHRONOLOGICAL ORDER

Nyayath Tharasu ('The Just Scale', 1989) revolves around the incidents in a two-week period, when a Naxal ('terrorist') activist is out on parole while serving a life sentence for the murder of a land owner, who she had seen kill a young tribal woman after she was raped and impregnated by him. The activist's mother, who used to be a freedom fighter, now on her deathbed, is relieved to see her daughter but is under the impression that she is free now. Her younger sister, her husband and her nephew are also happy to have her back home. But her younger brother, unemployed and an addict, is angered by her mere presence, blaming her for his inability to secure a good job. The mother dies. The activist's older brother who is home from Delhi to perform the death rites of their mother refuses to even talk to his activist sister, and after a big uproar leaves for his nephew to do the rites. In a hostile and intimidating context, the saving fact is that her old classmate who lives with her husband is close to her home.

A *Nyayath Tharasu* journalist tries to get an interview with the activist but she declines initially and is annoyed by his persistence. As the days pass, she feels unwanted, and ends up having no place to live, though now living with her sister. There is a suggestion of an affair with her brother-in-law. Then she asks the journalist for help and ends up staying at his place. A relationship of love blossoms between them and he succeeds in getting the government remission order in time, so that she no longer has to go back to jail. By this time there is some reconciliation in the family, and joy at her release. The 'activist' rushes to the place of her college mate to share the good news, but there she is shocked to see her being gang-raped by her husband and friends. True to her righteous self, the activist ends up shooting her friend's husband with his hunting rifle and ultimately surrenders herself at the police station. The story presents a complicated mixture of injustice, a faulty legal system, family attachments that could be good or bad, and finally multiple revenge as the solution.

Incidentally, *Meendum Parasakthi* ('Again Parasakthi', 1985) can confuse the public as its title corresponds to Karunanidhi's famous '*Parasakthi'*, 1952) but this is deceptive! Whatever was the motive, the film by a different screen writer focuses on the supreme power of the Goddess with reference to her myths and miracles. Combined with a love story and high musical score the film became popular. Finally, the Goddess's most ferocious form Badrakali reaffirms the popular imagination of her closeness to and reassurance for the ordinary masses.

Madurai Meenakshi (1993) by Karunanidhi appears to recall the message of his original *Parasakthi* of 1952. Since then there have been a number of films taken from stories (including the above) connected with Madurai, the *locus mundi* of Parasakthi, the Goddess or the Supreme Power. 'Madurai Meenakshi', though a deceptive title, touches aspects of political corruption and atrocity of a demoralised police force. 'Madurai' is a traffic policeman who lives with his mother and dreams of becoming a police inspector. One day, he clashes with Meenakshi, the daughter of a corrupt home minister. Meenakshi uses her father's power to change Madurai's duty, and Madurai becomes her house guard. Meenakshi spends her time humiliating the poor Madurai. At the same time his brother and his sister-in-law run a free school for blind children in their home. With crooked political manoeuvring the prime minister orders the home minister to buy their property. But they refuse to sell, and his henchmen kill Madurai's brother. In the meantime, Madurai and Meenakshi fall in love with each other. The home minister gets to know their love and transfers Madurai to the local police station. The Inspector and the home minister decide to buy the property by force and to put Madurai behind bars. (*A petition was filed against the film for negative portrayal of politicians, however the Supreme Court rejected the plea.*)

Pen Singam ('Lioness', 2010) focuses on the suffering of women by dowry and cheating. The stereotypes of corrupt politics, private industries that transgress the rules of public property such as forest areas, and gang violence are followed. An honest woman judge loses her life in the struggle. Her son, a forest officer and his lover who was persuaded and supported by him to become a police officer, choked by blood in encounter with gangs and police, seem to give hope for the future. It is insisted that there are lionesses among the lion family, but there can also be lions among women.

Ilaignan ('Young Man', 2011) is about a vibrant youth, the son of a liberation-minded woman, who wants to win the freedom of his people from a cruel and stone-hearted shipbuilding owner and his wife. With the help of the owner's sister, the young man starts a war against him. He and his wife play a cunning game and cheat the young man who is devoted to his mother. His mother is initially reluctant to wage war against evil, and wishes life for all. But the evil man crosses all the levels of cruelty and rejects the way of peace, so the mother asks her son to open a full-fledged war against evil, and in the end good wins over evil.

Karunanidhi's last film *Ponnar Shankar* (2011) was based on a historical episode. Though this and all his other films touch upon the themes of justice and faithfulness, both in terms of the title and his approach to evil in the society, they are stereotypes.

Climax of Intense Cruelty: *Puthiya Paraakthi* ('New Supreme Power', 1996)

Karunanidhi's opening speech recalls that he was a friend of MGR and he had penned three films starring him which made him famous. Then he mentions that there is no use in counting heads in democracy: only when we know what is inside those heads can we create true democracy and use it. Anna often used to say this. He used to illustrate it with the identification of an animal as an ox by a hundred people while twenty truly identified it as a cow but the latter would not win. Such a democracy based on untruth is taken by the dominant section with vested interest in the society and brainwashes the innocent.

Rajadurai, a village chief (President) is the embodiment of all cruelty, wickedness and womanising. Chinna Chamy works for him in many activities ranging from running a cinema to cleaning streets. His sweetheart Sakthi (power) or Parasakthi has as her family/personal deity the 'Parasakthi', 'supreme power'. Anyone who challenges Rajudurai is severely punished. He exploits the treasury of the temple, and when the priest points this out he is insulted by being made to ride facing backwards on a donkey. The police and some of the high officials are his blind supporters out of fear. Once Chinna Chamy was made the village President at Rajadurai's dictation but he exploited his power and signature while he was still working as his employee.

Chinna Chamy and Sakti's marriage is a significant development. There is a series of rapes by Rajadurai. Unobliging high officials are attacked. For forwarding a petition to the Collector, Chinna Chamy is severely beaten up. Also, in a false case of murder Chinna Chamy is sent to jail for six years. By the time of his return the couple has been blessed with a boy. On his return, the boy innocently intimates to his father: 'Mother says that there are those in jail who should be living outside and vice versa!' Chinna Chamy is silent while thinking.

Chinna Chamy exposes a grave injustice in the village. The people of the village were working in their own land but without knowing this, as there was no record in hand. Further reaction from Rajadurai sends him to the police lock-up under a false accusation. Chinna Chamy violently encounters some policemen and escapes. Sakthi is harassed with the allegation that her husband Chinna Chamy has stolen from Rajadurai's house. There is a retired 'military' uncle in the extended family who gives Sakthi and her son moral support. In the burning heat of cruelty and injustice, Sakti goes to the shrine of the deity, Parasakthi. She challenges her goddess with the words, 'Mother, have you been seeing all this cruelty?

It is said that God kills later, but in your name the whole village is punished in fire. If it is true that you destroy the wicked, then do they bribe you in the name of offertory and move around freely as though righteous? Are you testing us?' At the same time, the 'military uncle' makes efforts to recover the records of ownership of land for the people but he is killed, and his military uniform is inherited by the boy. Following the false implication of Chinna Chamy for two murders with people's false witness out of fear he too was murdered. Sakthi and her son go to the garden bungalow of Rajadurai.

Radadurai's serial rape now reaches his choicest victim Sakthi, but she and her son approach with a ploy. The boy sets fire to a haystack that creates confusion. At the height of his mother's struggle in the beastly rape the boy slips a sharpened axe to his mother through the window bars with which she attacks Rajadurai and kills him. When the police arrive the boy witnesses that his father Chinna Chamy came in the night and killed Rajaduraj just as he had come and murdered twice before (falsely implicated) and now, in the absence of Rajadurai, the people agree to witness with a signature or thumb impression.

The climax or anti-climax is that Sakthi is forced to follow the rituals of the tradition though she is a young widow whose husband has been cruelly killed. Finally, she foregoes flowers, colourful sarees and the auspicious dot mark on her forehead. The screen writer Karunanidhi speaks in the end: 'Just like we take out thorn by thorn, and cut diamond by diamond, we have to destroy the evil of the oppressive power of the dominant by the same method of evil. Here the mother Sakthi following the same method is a new Para Sakthi. It is like an antelope with her cub escaping a tiger.'

HIGHLIGHTS AND QUESTIONS

Coming from an ordinary family in a remote area, Karunanidh's arduous journey, achievements and fame were phenomenal. Screenwriter of about 70 films, author of about 100 books some of which won him prestigious awards, writer of his daily letter to his 'co-born' (party men), and mesmerizing orator in large crowds, he has influenced a wide range of people, friends and foes, across ideological camps and political allies. With a perennial eye problem, even while on a wheelchair for more than a decade, his discipline to write in beautiful Tamil (without having to delete or rewrite as he went!) has inspired many writers. His invariable appearance in white veshti and white shirt with a yellow scarf and dark glasses in middle class life-style established a unique identity of his personality. He has donated his main house in Chennai to be converted to be hospital. Even before and after his departure, his reminiscences have been propagated through mass media in order to keep his mark in history which may have had the effect of strengthening his party. Books on his life and work have started to appear in English as well and there is no doubt that the flow will continue (e.g., Ravishanker, 2019; Vaasanthi, 2020). However, in the light of our study of his films the following highlights and questions may stimulate further research with deeper analysis.

First, one may ask the fundamental question about his famous claim that he, following Periyar and Anna, truly represented the Dravidian vision and ethos with reference to the ancient Tamil literature, or helped to reconstruct it. What is unique in his stories as long as the Dravidian ethos is concerned? Is the Dravidian vision and polity an imaginative reconstruction of an exaggerated history? There is no indication about a historical deviation of a Dravidian past nor an alternative vision and a plan to rectify it.

Second, Karunanidhi has drawn the masses, but has he enlightened them? An empirical study of select individuals about their 'enlightenment' may be worthwhile if at all an open and unbiased account is possible. In any case, the masses he presents in his films are illiterate and superstitious. It is openly observed that the masses are taken as captivated audiences to participate in marches and conferences of the Dravidian parties with the inducements of free travel, food and allowances. In many films (e.g., *Neethikkuth Thandanai* and *Pensingam*) the comedians are projected as representing the dull-headed, pessimistic and ignorant average masses. If Karunanidhi expected that cinema would lead from entertainment to education and to enlightenment he must have been terribly frustrated. In *Puthiya Parasathi* he insists with an illustration of Anna that democracy succeeds not in the numbers game but when people internalise the facts and act with determination. Then, in the DMK elections, does democracy win through people of informed determination, or because of the influence of other forces and factors?

Third, when Karunanidhi presents social reality with inevitable twist and exaggeration he does not seem to have investigated what is inbuilt and happening in the human psyche. His favoured word is 'conscience' which is popular in Tamil parlance. *Neethikkuth Thandanai* talks about the death of conscience, and the twisting of it to the extent of blatant false witness by a person against his 'savior' in the court. The street singer tries to provoke people's conscience, but before the mighty nexus of politics, police and judiciary they remain weak. It is important to interrogate Karunanidhi about his view of the transformation of every person in and out with a view to create a society with justice and equality through a true democratic process.

Fourth, one may argue that, for true enlightenment and democracy, basically people need a decent social and material life. A sizable Tamil population has been reeling under poverty. The Anna government started to address this but it could not be sustained. Karunanidhi gave importance to infrastructure such as employment, roads, bridges and dams. But he was under pressure from the other governments to go for populous measures (freebies such as midday meals, televisions, wet-grinders and bicycles). These were seen as vote catchers too. There are other films which oppose this (e.g., *Valli*, 1993). But the notion and practice continue today with more offers such as subsidised public canteens. Also, the policies and schemes of the central government affecting the states and alliances add to the complexity.

Fifth, in almost all his films Karunanidhi highlights corruption, not only in government departments but also in the judiciary. Justice can be bought at different levels as we see in *Neethikkuth Thandanai*. Even in the last set of films including *Puthiya Parasakthi*, corruption from top to bottom creates bloodshed and murder. Most remarkably, seen from

Parasakthi (1952) the level of corruption in *Puthiya Paraakthi* (1996) has multiplied! What does this suggest? Even allowing for cinematic exaggeration, if it were not even half true the viewers would not believe the veracity of the story. Again, the important question is, as five-times Chief Minister, what attempt did Karunanidhi make to address the issue, and if so, what was the result? It has become a proverb recited in many films that at India's Independence Indian robbers replaced the Colonial robbers! Also, it has been pointed out publicly that the educated and those in position and power are more corrupt and exploiting than the uneducated and ordinary. What does this suggest?

Sixth, in Karunanidhi's social films violence has reached a phenomenal level. In *Parasakthi* (1952) there were a few events of physical violence – in the context of Gunanasekaran tackling the rowdies and attacking the priest who tried to get sexual pleasure with his sister. Kalyani was chased to the edge of life that caused her decision to kill her baby and commit suicide. Finally, the film has a positive ending with a programme of reconstruction. Multiple murders in *Neethikkuth Thandanai*, both public and private, send chill waves through average spines. As if justifying her murder and blaming the authorities, the heroine leads to declare a statement standing publicly in a bloodbath. *Puthiya Paraakthi* (1996) reaches the climax of many butchery murders. Now, can one imagine what had happened in the 44 years since the original *Parasakthi* (1952)? It is not only ironical but also shocking that he concludes this film with insisting 'evil for evil', throwing in the air the ideal of shaming evil by doing good as said by ancient sages including Thiruvalluvar. Incidentally, the celebrated film *Indian* of the same year (1996) with the famous actor Kamalhasan (b. 1954) as hero, as a former freedom fighter, exposes in a gruesome manner acts of corruption with murderous consequences. In a series of revenging murders, he finally kills his own son. As typical of most Tamil films, he rises to be a triumphant and invincible hero. The film sent waves of terror through the land, and the reason for its success was said to be in publicising the victims of corruption. Kamalhasan, who acknowledges that in his cinema career he was influenced by Karunanidhi's dialogues, seems to follow the principle of 'evil for evil' in many of his other films as well. At the same time there is no dearth of ideas and ethical teachings in some of his films and speeches. Most remarkable is the film *Anbe Sivam* (God or Siva is Love, 2003) in which he starred as the main hero, and in which there is a touching demonstration of love in helping people in crisis and trying to overcome evil by good. And in connection with different views of tackling the 2020 Coronavirus pandemic his public advice was 'Help one another, and love and knowledge have no end' (*Daily Thanthi,* 2020). It is a relevant question whether the pedagogy of the Tamil cinema can be geared to this approach in the place of the overarching principle of evil for evil. He established his own political party in Feb 2018, to add to the mushrooming number of parties established, some by former film actors, with the title 'Centre for People's Justice' (*Makkal Neethi Maiyyam*). Now while his 'Indian-2' is in the making there is no sign of providing an alternative vision, ethos and method. Moreover, has Karunanidhi ever applied his mentor Anna's famous words 'Do not sharpen your knife but your intellect'? Could he not find an alternative way of analysing with vision from the Dravidian ethos of civility where truth-telling and truth-listening is possible. The greatness of the Dravidian civilization of both the Indus Valley and ancient Tamils cannot lie in historical antiquity but in the level of thinking and the quality of life today. Otherwise, a true judgment is that the Dravidian civilization is a construct of false imagination barring scattered ideas of individual and moral teaching.

Seventh, on the one hand, as a Dravidian ethos, female sexual chastity is divinised. Affection for mother (divinity attached) and sister is highly celebrated. Karunanidhi's family films elaborate on this, and any violation is duly punished, as is evident in *Parasakthi.* On the other hand, in subsequent films sexual violence is projected in a most obscene yet cruel fashion. In *Neethikkuth Thandanai* the violent rape scenes involving Chitra with the police officer Chandrakandh and minister Markendeyan may expose the reality of innocent women being subjected to beastly rapes. At the same time, is there a perverted sense of entertainment as well – preying on rape by violence? In any case, during her second encounter with Markendeyan, Chitra (now in the guise of Sinduja) tells him 'I am no more a coward but Badrakali (the ferocious goddess of justice)', and these words may empower potential victims. But is Chitra's disguise as Sinduja viable for ordinary women? Can there be an alternative approach to the problem?

Eighth, fame seems to be a peculiar Dravidian ideal. According to Thiruvalluvar, 'If born, be born with fame; or else, better not be born than born.' Karunanidhi's fame is obvious. When his political rival Jayalalitha died he paid a generous tribute acknowledging her achievements and forecasting that her fame would remain for many years. And the personality cult has emerged as a characteristic mark of the Dravidian parties. The number of public statues can be taken to assess the magnitude of fame. At the climax Karunanidhi decided to allot a prime spot on Marina Beach in Chennai for the burial of his 'heart's deity' (Anna) and by the same logic (dying in service as Chief Minister) MGR and Jayalalitha were qualified to be buried there. But when Karunanidhi died he was no longer Chief Minister and was denied a burial place near his mentor. A night's legal battle in the court cleared the space, and the recognition was for his fame. The whole episode seems to expose the nexus between piety, power and politics in Tamil Nadu (Selvanayagam 2019, 1-18). Is there an ideal of a collective fame in the Dravidian tradition? If there is none, nothing can prevent the present generation from developing it by an enlightened community!

Ninth, as we pointed out at the beginning, Periyar was a hard core atheist. Anna was a little more open with his watchword 'One Clan (*humanity*), One God'. Karunanidhi used to claim that he was an avowed atheist. But as religion was part of common people's life, most of his film stories have religions, deities and festivals. But he was nuanced when he chose the word *Parasakthi* ('Supreme Power') for his most decisive film in 1952. There he did not ridicule religion but its misunderstanding and misuse. Some may see that his differing views of religion (e.g., the godman being silent in *Palaivana Rojakkal*, 1986) were informed by political trends with Brahmanic influence. In *Puthiya Parasakthi* once again, but in a violent manner, the Goddess is challenged for supporting the wicked for their bribes in the form offertory and justifying the cruelty of the wicked. In the meantime, when Jayalalitha became Chief Minister one of her first actions was to renovate temples and promote Vedic study. One could see the ploy that the Brahmanic forces which were the arch-enemies of the Dravidian movement in their typical manner throughout history found a political space to use MGR and Jayalalitha and reinforce their power. Consequently, today the ruling faction with the names 'Dravidian' and 'Anna' in their name (AIADMK), with support from the Bharatiya Janata Party government at the centre which is openly engaged in revitalising Brahmanic Hindu tradition, has rituals performed by Brahmanic priests at the opening of every government and public work function! It is unbelievable that the other secular parties and leaders keep quiet. There are few greater ironies in history, and the bodies of Periyar and Anna may be rolling in their graves in agony! Moreover, apart from Brahmanic Hinduism and its 'superstitious rituals' (belief in ritual's self-generating power when handled by the Brahmin priests that succours the gods, sustains the universe and maintains the caste-based social order) there are a number of religious traditions in Tamil Nadu. Karunanidhi has not made any effort to find out the liberative visions and traditions of certain religious traditions even if they are either confused by Brahmanic beliefs or have deviated from their original vision.

Tenth and final, there are two strong factions of the Dravidian movement with their allies based in Chennai. Their identities are ascertained by two portraits kept on walls and behind desks. The ruling AIADMK holds the portrait of Jayalalitha, an actor turned politician, and the DMK holds the portrait of Karunanidhi, a patriarch of the Dravidian movement, writer, speaker, playwright and politician. Their films will be played out again in the next election. But badly needed is a serious review, recognising their ideological bankruptcy (self-dignity?), and exposing their methodological flaws. Whether a third wave will emerge in the light of some of the recent films (mobilization of masses with focus on specific issues) and establish the true Dravidian civilization or common human civilization and ideals, we must wait and see!

REFERENCES

Baskaran, Theodore, 2020, *Thandorak-kaararkal* (Tami, 'Theatrical Drummers'), Vanthavasi: Akani Publications.

Daily Thanthi E-paper, 24 April 2020.

Diehl, Anita 1978, *Periyar EVR: A Study of the Influence of a Personality in Contemporary South India*, New Delhi: B.I. Publications

Gita, V., 2006 "Views of Western Scholars", in *Periyar: Dravidam & Indhuthuvam* (Tamil, Dravianism and Hindutva), Essays compiled by Sahayaraj S.J., Dindigul: Vaigarai Publishing House, 79-88.

Kannan, R., 2010 *Anna: The Life and Times of C.N. Annadurai*, Delhi: Penguin Books India.

Kannan, R. 2017, *MGR: A Life,* Delhi: Penguin Books India.

Ravishankar, Sandhya 2020, *Karunanidhi: A Life in Politics,* London: Fourth Estate.

Ryerson, 1988, Charles Ryerson, *Regionalism and Religion: The Tamil Renaissance and Popular Hinduism,* Chennai: Christian Literature Society (for the Christian Institute for the Study of Religion and Society, Bangalore), 1988, 105.

Selvanayagam, Israel, "The Nexus of Piety, Power and Politics in Tamilnadu: A View through the Lens of Four Tombs in the Capital City Chennai." *Black Theology.* Volume 17, Issue 2 (https://doi.org/10.1080/14769948.2019.1627092) 18 June 2019, 1-18.

Vaasanthi 2016, *Amma: Jayalalitha's Journey from Movie Star to Political Queen*, New Delhi: Juggernaut.

Vaasanthi 2020, *Karunanidhi: The Definitive Biography*, New Delhi: Juggernaut.

Website 2020, *(For some important snapshots of conflict between Karunanidhi and Jayalalitha)* https://www.google.co.uk/search?source=hp&ei=jNyaXp-iMoi8Uv-MqfAL&q=karunanidhi+jayalalitha&oq=Karunanidhi-Jayalalith&gs_lcp=CgZwc3ktYWIQARgAMgQIABAeMgQIABAeMgQIABAeMgQIABAeMgYIABAKEB4yBAgAEB4yBAgAEB4yBggAEAgQHjIGCAAQCBAeMgYIABAIEB46AggAOgUIABCDAToGCAAQDRAeUL0RWINyYOHCAWgAcAB4AIABWIgB-QuSAQIyMpgBAKABAaoBB2d3cy13aXo&sclient=psy-ab – as on 18 April 2020.

In: India
Editors: J. Lessard and O. Rivière
ISBN: 978-1-53618-413-6

Chapter 5

SOCIO-ECONOMIC FACTORS AFFECTING CHILD MALNUTRITION IN UTTAR PRADESH, INDIA

Reena Kumari[1,*] and Awaneendra Kumar Tiwari[2,†]
[1]Department of Economics, Chandradhari Mithila Mahavidyalaya, L.N. Mithila University, Darbhanga, Bihar, India
[2]Research Scholar, Department of Electrical Engineering, IIT(ISM), Dhanbad, Jharkhand, India

ABSTRACT

This chapter examines the effects of socio-economic determinants on child malnutrition in Uttar Pradesh- the most populous and malnourished state of India. Using data from NFHS-4, this study highlights district-wise variations in nutritional status of children less than five years in the state. The percentage of stunted, wasted and underweight have been taken as dependent variables, while main explanatory variables of child malnutrition were female education, maternal malnutrition (BMI),

* Corresponding Author's E-mail: raireena86@gmail.com.
† Corresponding Author's E-mail: awi.tiwari@gmail.com.

breastfeeding practices, children's place of residence, father's education, wealth index of household, family planning, sex of the child, sex of the household and age cohort of mother. Logistic regression models have been applied in order to observe the association between explanatory variables and malnutrition. The findings revealed a high incidence of less stunted, wasted and underweight infants in case of an educated parents, particularly educational level of mother. Wealth index representing the possession of assets by the household found to be an important indicator in the sense that children born to households with richer and richest quintiles showing better-nourished children aged under-5. The modern method of using contraceptive, breastfeeding practices, maternal malnutrition were also important indicators influencing the rate of malnutrition in children. The study suggested that imparting education to females, breastfeeding practices, economic status and contraceptive use can have an important role in reducing nutritional status of children.

Keywords: malnutrition, stunted, wealth index, education, contraceptive use

INTRODUCTION

Malnutrition, defined as deficiency or imbalances in a person's nutrients intake, is considered a sensitive indicator and most common nutritional disorder for any developing nation. The high prevalence of child malnutrition reflects not only a country's past health status but also its future health trails (Subramanyam et al., 2010). Addressing the problem of malnutrition, the sustainable development goals (SDG) adopted by United Nations (2019) indicates *'By 2030, end all forms of malnutrition, including achieving, by 2025, the internationally agreed targets on stunting and wasting in children under 5 years of age, and address the nutritional needs of adolescent girls, pregnant and lactating women and older persons'* under Goal 2. In this context, monitoring child malnutrition is crucial for planners and policy makers in a country like India, a country which is facing the high prevalence of child malnutrition, accounting for one third of the world's wasted (height-for-age). UNICEF contends that in India, the percentage of underweight and stunted children less than five years were 43 percent and 48 percent respectively because of enduring undernutrition. It is estimated that worldwide, India executes poorly in child malnutrition status and ranked 114 out of 132 countries, just prior to Afghanistan and Pakistan (IFPRI, 2016) and in the early 2000s, it was around one-third of all malnourished children in the world (Svedberg, 2008). Several studies indicated that child malnutrition affected child morbidity, health-care expenditure and economic progress of any country (Measham & Chattergee (1999); Mishra, Lahiri & Luther, 1999) which further increase the risk of child ailment, retardation in mental and physical growth, resulting into low ability to work, loss of output and efficiency, high burden of health care expenditures and lastly can pledge the poverty trap (World Bank, 2006; Grantham et al., 2007; Tarozzi & Mahajan, 2007; Walker et al., 2007).

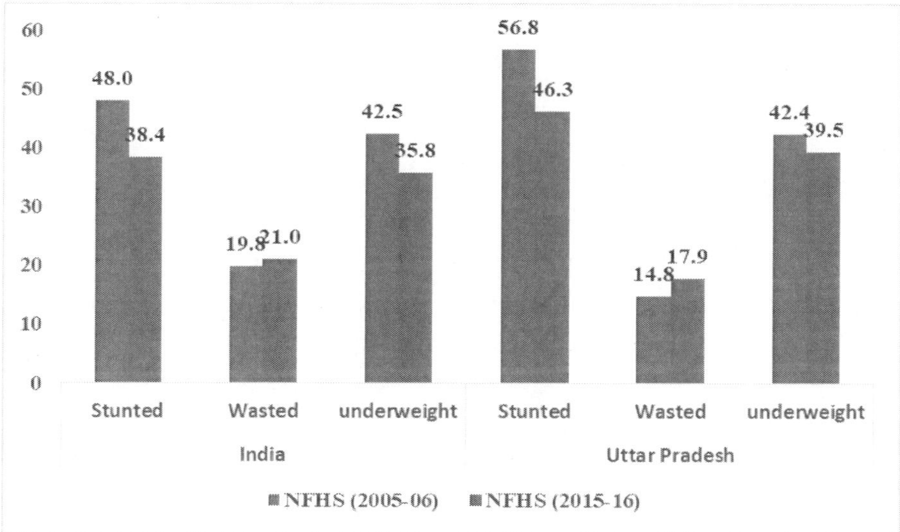

Figure 1. Malnutrition trends in India and Uttar Pradesh children among aged 0-5 years.

Also. It is considered that severe or chronic malnutrition increase the educational attainment and outcome of the children (Ahmed et al., 2012). International Food Policy Research Institute (IFPRI) reports that in developing countries, one out of every three children under-five, is malnourished. It is clearly noted that the high prevalence of malnutrition not only cause severe cognitive and physical loss of children but also it is a collapse or exploitation of a child's human right (Das & Sahoo, 2011). The population of Uttar Pradesh is 199.5 million (India's most populated state) and 39 percent of the total child population of India resides in the state. According to the report published by Uttar Pradesh State Nutrition Mission, in this state, half of all children age-5 are stunted and 10 percent were wasted in 2014. In Uttar Pradesh, infant mortality rate (IMR) and malnutrition rate are the highest in the country. National Family Health Survey-4 (NFHS) data revealed that in Uttar Pradesh, the percentage of malnourished children has declined significantly in the last ten years. In 2005-06, 56.8 percent of children under-5 years of age were stunted and 42.4 percent were underweight; these figures fell to 46.3 percent and 39.5 percent respectively, in 2015-16 (Figure 1). As per the report published by National Sample Survey Office (NSSO), the percentage of poor in the state is 30.4 percent higher than the all India (25.7%) poor. IMR is 48 per thousand live births, the under-5 mortality rate is 73 per thousand live births, and life expectancy is 62.9 years for male and 65.4 years for female. However, the mortality rate among the children under-5 years of age fell by 18 percent from 2005-06 to 2015-16, and IMR fell by 9 percent during the same period. Improvement in child nutrition during the inter-survey period is the result of interventions such as broader coverage of midday meals to children at school and rigorous operation of Integrated Child Development Scheme (ICDS) at a more disaggregated level, underpinning of health outreach facilities to rural and poor ménages for on-time vaccination of children and strengthening of the public distribution system (PDS) of providing food grains at promoted prices to the weaker section. However, undernutrition of children still remains a major health problem at the disaggregated level in India. There has been a remarkable difference in IMR across the districts; for example, highest in Shrawasti (96) while lowest in Kanpur Dehat (37). In Uttar Pradesh, a considerable variation has been observed between rural and urban in terms of child malnutrition. In rural areas of Uttar Pradesh, prevalence of stunted and underweight among children was 48.5 percent and 41 percent respectively, as compared with

37.9 percent and 33.7 percent among urban children. In this framework, a study of child malnutrition and its important determinants is required for policy implications and therefore, the present study examined the effects of various socio-economic factors on child malnutrition children less than five years in Uttar Pradesh, employing the data from NFHS-4 (2015-16).

PREVIOUS RESEARCH

There are several socio-economic and demographic factors (place of residence, age and nutritional status; and maternal education) responsible for poor nutritional status of children (Bangladesh Demographic and Health Survey, 2007). The benefits of mother's education and its impact on child health status is well documented in literature which lowers the rate of child malnutrition, in other words, the prevalence of malnutrition is lower in case of educated mother (Frost, Forste & Hass, 2005; Basu & Stephenson, 2005; Miller & Rodgers, 2009; Cleland, 2010; Bbaale, 2014; Burroway, 2016). A study by Mosle & Chen (1985) reported that effect of mother's education on children's nutritional status works through a set of 'proximate determinants,' such as, fertility factors, environmental amenities, breastfeeding practices and utilization of childcare services. Similarly, a host of studies contend that more educated mothers may have more nourished children because they have better knowledge about child rearing practices and heath care and utilization (Thomas et al., 1991; Desai & Alva, 1998; Glewwe 1999; Currie & Moretti, 2003). A substantial range of literature in the domain of health and education have discussed the casual relationship between mother's education and nutritional status of children and given some possible trails to found the link. The important studies include: (1) more educated mothers can take better care of their children about nutrition and health care, for example, washing of clothes, feeding practices, treatment during sickness and availing timely vaccination services etc. (Caldwell 1979; Mondal et al., 2009) and better sanitation such as flushing toilets and running water (Horton, 1998; Case, 2001, Choudhury, 2015) (2) An educated mother can make healthier choices for themselves during pregnancy that effectively prevent fatal childhood diseases (Govindasamy & Ramesh, 1997) and directly affected the health of the child at birth (Miller & Rodgers, 2009); (3) Educated mother can give better level of living as previous research documented that education provides women ample opportunities to work outside the home and earn an income which empower them through greater authority and bargaining capability in the family (Nussbaum, 2004; Sen, 1999).

However, a growing body of literature has drawn attention towards the impact of father's education in improving child nutritional status (Case et al., 2002). It is well-established fact now that educational level of father plays a crucial role in family's financial status which in turn ensures better access to child health facilities and further better child nutritional status (Choudhury, 2015). India's NFHS data also revealed a very high percentage of maternal malnutrition which shows the worse-off condition of women. Studies claimed that mother's nutritional status had a positive association with child nutritional status (Sethuraman, 2006).

Poor breast-feeding practices in early childhood contribute to the burden of malnutrition and infant and child mortality (Patel et al., 2010). It is observed that working mothers have limited time for breastfeeding and also to manage other nutritional and health services to their children. Working mothers are dependent on a caretaker for their child and caretaker's poor attention, affection and involvement in child feeding are the major causes for undernutrition of children. Moreover, a large body of empirical research claimed that caretakers might not make the best use of offered resources due to less information of optimal feeding behaviours and inappropriate cultural beliefs and practice about nourishing (Allen and Gillespie, 2001; Moore, Akhter and Aboud, 2006).

Also, location or place of residence is one of the socioeconomic covariates often used in the literature which determines child malnutrition across the region (Sastry, 1997). A growing body of literature demonstrated that wealth status of a family has a significant and positive impact on nutritional status of children (Smith & Haddad, 2000; Haddad et al., 2003; Heltberg, 2009; Headey, 2013). A variety of indicators has been used to measure economic status of the household, however, sorting of households according to income level is more relevant proxy. While many studies claimed that malnutrition is associated with poverty and disease (Dasgupta et al., 2005) and argued that due to poor access of food a large segment of children is suffering from severe health problems and morbidity. To improve child nutritional status, Government of India has introduced major nutrition supplementation programs such as Integrated Child Development Services (ICDS) and Mid-day Meal (MDM) Programme at the disaggregated level. However, little is known about the empirical findings of the factors that drive child malnutrition, especially in poor and bigger states like Uttar Pradesh. In this connection, in order to formulate appropriate strategies related to child health outcomes, policy makers and planners must prerequisite familiarity of the factors responsible for poor child outcome such as stunting, wasting and underweight. Therefore, the main objectives of this chapter are to understand nutritional status based on stunted, wasted and underweight of children less than the age of five and to examine the factors which influence child malnutrition in respect of certain socio-economic variables, for instance, father's education, place of residence, wealth status, family planning, sex of the child, sex of the household, religion; and maternal indicators like mother's malnutrition (BMI) and age cohort of mother. This study examined the various socio-economic and household factors which are responsible for child malnutrition in Uttar Pradesh.

MATERIAL AND METHODS

In this study, data has been assessed from the NFHS-4 conducted by the International Institute of Population Sciences (IIPS), Mumbai in 2015-16 and constrained to children under the age of five. NFHS is a collaborative project of IIPS and many organizations including foreign institutions.

Table 1. Symbolization and definition of variables used in the logistic regression

Symbolization	Independent variables	Definition
Mother's education	Educational level of the mother (dummy variable)	
Mother_illit[R]	Mother is illiterate	=1, if mother is illiterate =0, otherwise
Mother_elemedn	Mother received elementary level of education	=1, if mother is having 1-8 years of schooling =0, otherwise
Mother_secedn	Mother received secondary level of education	=1, if mother is having 9-10 years of schooling =0, otherwise
Mother_hiedn	Mother received senior secondary or above level of education	=1, if mother is having 11 or more years of schooling =0, otherwise
Sex_child	Sex of child (dummy variable)	
Child_boy[R]	Boy child	=1, if the child is boy =0, otherwise
Child_girl	Girl child	=1, if the child is girl =0, otherwise
Father's education	Educational level of the father (dummy variable)	
Father_illiterate[R]	Father is illiterate	=1, if father is illiterate =0, otherwise
Father_elemedn	Father received elementary level of education	=1, if father is having 1-8 years of schooling =0, otherwise
Father_secedn	Father received secondary level of education	=1, if father is having 9-10 years of schooling =0, otherwise
Father_hiedn	Father received senior secondary or above level of education	=1, if father is having 11 or more years of schooling =0, otherwise
Sex_hhn	Sex of Household (Dummy variable)	
Male[R]	Male member	=1, if the sex of household is male =0, otherwise
Female	Female member	1, if the sex of household is female =0, otherwise
Place_residence	Place of residence (dummy variable)	
Urban[R]	Place of residence in urban areas	=1, if the location of household is urban =0, otherwise

Symbolization	**Independent variables**	**Definition**
Rural	Place of residence in rural areas	=1, if the location of household is rural =0, otherwise
Contraceptive_use	Use of contraceptive or family planning by household (dummy variable)	
Trad_method^{R}	Household using traditional method	=1, if the household is using traditional method =0, otherwise
Mode_method	Household using modern method	=1, if the household is using modern method =0, otherwise
Others_method	Household using other method	=1, if the household is using other method excluding traditional and modern =0, otherwise
Wealth Index	Wealth index reflecting the ownership of some basic assets by the household (dummy variable)	
Poorest^{R}	Household is belonging to the poorest income category	=1, if the household is belonging to the poorest category =0, otherwise
Poor	Household is poor	1, if the household is poor =0, otherwise
Middle	Household is belonging to the middle-income category	1, if the household is belonging to the middle-income category =0, otherwise
Richer	Household is belonging to the richer income category	1, if the household is belonging to the richer income category =0, otherwise
Richest	Household is belonging to the richest income category	1, if the household belonging to the richest income category =0, otherwise
Body Mass Index (BMI) of mother	BMI of the mother (dummy variable)	
BMI ≤18.5 kg/m2R	Mother's BMI is BMI ≤18.5 kg/m2	=1, if the mother's BMI is BMI ≤18.5 kg/m2 =0, otherwise
BMI ≥18.5 kg/m2	Mother's BMI is BMI ≥18.5 kg/m2	=1, if the mother's BMI is BMI BMI ≥18.5 kg/m2 =0, otherwise
Breastfeeding	Breastfeeding practices by mother (dummy variable)	
Within 1st hourR	Child received breastmilk within 1st hour	=1, if child received breastmilk within 1st hour =0, otherwise

Table 1. (Continued)

Symbolization	Independent variables	Definition
More than 1st hour	Child received breastmilk more than 1st hour	=1, if child received breastmilk more than 1st hour =0, otherwise
Mother's Age Cohort	Age of the mother (dummy variable)	
15-19R	Mother is 15-20 years old	=1, if the mother is 15-20 years old =0, otherwise
20-24	Mother is 20-24 years old	=1, if the mother is 20-24 years old =0, otherwise
25-29	Mother is 25-29 years old	=1, if the mother is 25-29 years old =0, otherwise
30-34	Mother is 30-34 years old	=1, if the mother is 30-34 years old =0, otherwise
35-39	Mother is 35-39 years old	=1, if the mother is 35-39 years old =0, otherwise
40-44	Mother is 40-44 years old	=1, if the mother is 40-44 years old =0, otherwise
45-49	Mother is 45-49 years old	=1, if the mother is 45-49 years old =0, otherwise

IIPS was assigned as the nodal agency by the Ministry of Health and Family Welfare, Government of India and responsible for providing coordination and technical assistance for the NFHS-4[1]. The survey used a uniform questionnaire, two-stage sampling design with stratification by rural-urban background and field practice to make possible comparability of the data and to achieve enhanced data quality. For anthropometric analysis, malnutrition of children is enumerated in terms of anthropometrical measures- weight-for-age (stunted), height-for-age (wasted) and weight-for-height (underweight) and Z-score ≤ 2 standard deviation, World Health Organisation/National Centre for Health Statistics reference standards were applied to compute the standardized Z-score for all three measures (WHO, 1978; Mazumdar, 2010).

[1] NFHS-4 is funded by the United States Agency for International Development (USAID), DFID, the Bill and Melinda Gates Foundation (BMGF), UNICEF, UNFPA, the MacArthur Foundation and the Ministry of Health and Family Welfare (MoHFW), Government of India.

In this survey, data on children's height, weight and underweight have been collected less than 5 years. Mother's education, place of residence (rural-urban), sex of the child (boys-girls), sex of the household, father's education, wealth index (based on an ownership-of-goods index), family planning (use of contraceptive), mother's BMI, breastfeeding practices and mother's age cohort are the important indicators that are assumed to effect nutritional status of a child. The logistic regression analysis has been employed in the analysis for estimation of the odds of being malnourished in the children of Uttar Pradesh. Dependent variable denotes the number of children whose z-scores are below -2 are coded 1 and those with z-scores of -2 or higher are coded 0. The details of explanatory variables have been given in Table 1 and thus results obtained are compared with reference category. In order to compute variables, research design and sample weights, statistical tool is carried out using SPSS (20.0 version) software for window and the significance levels of $p < 0.01$, 0.05 and 0.10 were taken.

RESULTS AND DISCUSSION

Table 2 clearly depicted that the percentage of underweight and stunted children increased up to one year for the boys and then decreased to about 4 years. The percentage of wasted children increased up to one year and then decreased up to 4 years for both, the boys and girls. The data demonstrates that the percentage of underweight, stunted and wasted below the age of five was 39.7 percent, 46.5 percent and 19.2 percent, respectively for boys and it was 39.8 percent, 46.2 percent and 16.7 percent for girls. The data also revealed that the percentage of underweight was less in girls in comparison to boys, but it is in contrast to the case of stunted and wasted.

District-wise variations in frequencies and ranks among the under-five children by sex were given in appendix (Table 1A). It is clearly noted that inter-district variation is high in terms of stunted, wasted and underweight in Uttar Pradesh. In terms of boys stunted, the top five poor performer districts were Gautam Buddha Nagar, Jhansi, Ballia, Meerut and Saharanpur. While, all the bottom five performer districts where the prevalence of stunting was high were; Bahraich, Shrawasti, Balrampur, Gonda and Siddharth Nagar, located in the eastern region of Uttar Pradesh. Despite being placed in the western region in the state, Gautam Buddha Nagar and Saharanpur, both have a high percentage of stunted children. In terms of boys wasted, Mahamaya Nagar, Etah, Farrukhabad, Gonda and Sant Kabir Nagar were the top five performers, while Hamirpur, Lucknow, Lalitpur, Kaushambi and Jalaun were the bottom five performers. The districts which fall under the top five performers in boys underweight include Mathura, Mahamaya Nagar, Gorakhpur, Gautam Buddha Nagar and Firozabad. In contrast, Budaun, Kaushambi, Chitrakoot, Jaunpur and Pilibhit were the bottom five performers.

In terms of girls stunted, Ghaziabad, Baghpat, Lucknow, Gautam Buddha Nagar and Rae Bareli were the top five performers while, Bahraich, Balrampur, Shrawasti, Siddharth Nagar and Maharajganj were the poor performers. In girls wasted category, Lalitpur, Chitrakoot, Rae Bareli, Lucknow and Hamirpur were the bottom five performers while, Mainpuri, Farrukhabad, Kanshiram Nagar, Gonda and Deoria were the top five performers. In terms of girls underweight, districts fall into the five top categories were Firozabad, Mathura, Ghaziabad, Ballia and Gautam Buddha Nagar while, Shahjahanpur, Jaunpur, Budaun, Jalaun and Sant Ravidas Nagar were the bottom five performers in the state.

Districts with a high prevalence of stunted for both the boys and girls were observed in Bahraich, Balrampur, Shrawasti, Siddharth Nagar while the low prevalence for both sexes was observed in Gautam Buddha Nagar. Likewise, in terms of wasted, the high prevalence for both sexes has been observed in some districts like Hamirpur, Lalitpur and Lucknow. In contrast, the low prevalence for boys and girls has been observed in Gonda. High prevalence of underweight for both boys and girls has been observed in Budaun. In contrast, low prevalence of underweight for both has been seen in the districts of Mathura, Forozabad and Gautam Buddha Nagar.

The percentage of stunted is higher in girls in comparison to boys in the districts like, Rampur, Shahjahanpur, Unnao, Jalaun, Fatehpur, Basti, Ballia, Chandauli and Etah. Some districts in the state where the percentage of stunted was lower in girls in comparison to their boy's counterpart include Jyoptiba Phule Nagar, Firozabad, Pilibhit, Hardoi, Lalitpur, Faizabad, Ambedkar Nagar, Jaunpur, Ghazipur and Sonbhadra. No substantial differences between boys and girls were observed in case of wasting. Although, considerable variation between boys and girls were found in some districts such as Saharanpur, Mahamaya Nagar, Bareily, Hamirpur, Fatehpur, Balrampur, Sant Kabir Nagar and Gorakhpur where the prevalence of underweight was higher in boys than girls. On the other hand, there were some districts where the percentage of underweight was higher in boys in comparison to girls include Mainpuri, Pilibhit, Kanpur Nagar, Kanpur Dehat, Barabanki, Sultanpur, Bahraich, Maharajganj and Jaunpur.

Table 3 reveals the parameter estimates for stunting. Coefficient of covariates and the related standard error is shown in parentheses. Mother's level of education is an important factor in determining stunting of a child. It was found that mothers who received elementary (OD = 1.598, SE = .129), secondary (OD = 1.762, SE = .137) and higher (OD = 1.442, SE = .117) level of education lower the rates of stunting. This finding is unfailing and supported by many studies such as Roy (2000); Smith & Hadded (2000); Headey (2013) and many others.

Likewise, father's education also has a significant influence on stunting. The result indicates that fathers who have completed elementary (OD = 1.347, SE = .125) and secondary (OD = 1.451, SE = .128) level of education condense the rate of stunting and it is statistically significant at 1 percent level of significance. Though, the result is not found to be statistically significant if father has above secondary or higher level of education.

However, household's location of children no longer affected the rate of stunting. The association between sex of household and stunting is negative and statistically significant at 1 percent level of significance with low odd ratio. It confirms that if the of household is headed by female (OD = .760, SE = .082) it lowers the rate of stunting.

Another important and determining factor of stunting is wealth index. It demonstrates that children born to families with richer and richest wealth quintiles showing lower stunting children. It means if the economic condition of a family is better it can upsurge the level of living of the household, allow them to take indispensable care of the children and further reduce stunting of children.

Table 2. Age and sex wise nutritional status of under-five children in Uttar Pradesh

Sex of child	Year	Stunted		Wasted		Underweight	
		N	%	N	%	N	%
Boy	0	914	24.9	1113	30.3	1185	32.3
	1	2032	53.2	826	21.6	1618	42.4
	2	1982	52.9	629	16.8	1573	42.0
	3	2046	52.5	588	15.1	1611	41.4
	4	1860	48.1	499	12.9	1557	40.3
	Total	8834	46.5	3655	19.2	7544	39.7
Girl	0	724	21.7	955	28.6	1032	30.9
	1	1659	48.1	642	18.6	1322	38.3
	2	1889	55.3	460	13.5	1467	43.0
	3	1970	54.0	415	11.4	1562	42.8
	4	1719	51.1	396	11.8	1468	43.6
	Total	7961	46.2	2868	16.7	6851	39.8

Source: Calculated from NFHS-4 data, Government of India.

Use of modern method of contraceptive is significant indicator which also lowers the rate of stunting. The results display that if family is using modern method (OD=1.291, SE=.255) of contraceptive, they reduce stunting.

The association between mother's BMI (OD = 2.654, SE = .082) and stunting is robust and highly significant with high odd ratio. The result is statistically significant at 1 percent level of significance which indicates if mother's nutritional status is better, they can lower the rate of stunting.

Our logistic regression results indicate a significant and negative relationship between children belonging to the age up to 5 years who have never been breastfed within first hour of birth (OD = .876, SE = .063) and stunting. There is no significance association found between mother's age cohort and stunting.

Table 4 represents that mother's education has a robust and positive effect on wasting relative to mothers with no education (model 1). The odds ratios for elementary (OD = 1.633, SE = .127), secondary (OD = 1.781, SE = .135) and higher level of education (OD = 1.425, SE = .116) are more than one for all levels of education.

Table 3. Determinants of stunting using multiple logistic regression results

Variable	B	Sig.	Odd ratio	95% C.I.
Mother's Education				
Mother_illit[R]				
Mother_elemedn	.469 (.129)	.000	1.598	1.241-2.058
Mother_secedn	.567 (.137)	.000	1.762	1.347-2.306
Mother_hiedn	.366 (.117)	.002	1.442	1.147-1.814
Sex of Child				
Child_boy[R]				
Child_girl	.093 (.056)	.099	1.097	.983-1.225
Sex of Household				
Male[R]				
Female	-.275 (.082)	.001	.760	.646-.893
Father's Education				
Father_illiterate[R]				
Father_elemedn	.298 (.125)	.017	1.347	1.054-1.720
Father_secedn	.373 (.128)	.004	1.451	1.130-1.864
Father_hiedn	.101 (.104)	.331	1.106	.903-1.356
Place of Residence				
Rural	.132 (.079)	.095	1.142	.977-1.333
Use of Contraceptive				
Trad_method[R]	.089 (.101)	.380	1.093	.896-1.332
Mode_method	.255 (.110)	.021	1.291	1.040-1.602
Others_method	.064 (.092)	.489	1.066	.890-1.276
Wealth Index				
Poorest[R]				
Poor	1.121 (.130)	.000	3.069	2.379-3.960
Middle	.866 (.124)	.000	2.377	1.865-3.030
Richer	.593 (.122)	.000	1.810	1.424-2.300
Richest	.366 (.124)	.003	1.442	1.131-1.840

Table 3. (Continued)

Variable	B	Sig.	Odd ratio	95% C.I.
Body Mass Index (BMI)				
BMI ≤18.5 kg/m2[R]				
BMI ≥18.5 kg/m2	.976 (.082)	.000	2.654	2.260-3.116
Breastfeeding				
Within 1st hour[R]				
More than 1st hour	-.132 (.063)	.037	.876	.774-.992
Mother's Age Cohort				
15-19[R]				
20-24	-.524 (.460)	.255	.592	.241-1.459
25-29	-.263 (.369)	.476	.769	.373-1.585
30-34	-.048 (.367)	.896	.953	.464-1.957
35-39	.008 (.369)	.983	1.008	.489-2.075
40-44	-.070 (.375)	.852	.933	.447-1.947
45-49	.011 (.403)	.979	1.011	.458-2.229
N	41751			
Cox & Snell R Square	0.085			
Nagelkerke R Square	0.113			
-2 Log-likelihood	7608.059			

R, reference category.
Figures in the parentheses are SE of estimates.

Father's educational status also has a significant effect on wasting. The prevalence of wasting is lower in the children whose father's level of education is elementary (OD = 1.364, SE = .123) and secondary (OD = 1.465, SE = .126). However, father's education above secondary level is not a necessary condition for wasting.

The prevalence of wasting is higher in children who are residing in rural areas. The result demonstrates a significant effect on wasting with high odd ratio (OD = 1.168, SE = .078). As expected, due to poverty, rural people have low access to the health care facility and better sanitation that result in a high percentage of wasting.

Also, sex of the household is another variable which determines wasting. The prevalence of wasting is higher in the households which is headed by female (OD = .759, SE = .081). The result is statistically significant at 1 percent level of significance. Similar to this result, Burroway (2016) revealed an important finding that females are more efficient than their male counterpart to use resources on nourishment and other basic domestic needs.

Table 4. Determinants of wasting using multiple logistic regression results

Variable	β	Sig.	Odd ratio	95% C.I.
Mother's Education				
Mother_illit^R				
Mother_elemedn	.490 (.127)	.000	1.633	1.273-2.094
Mother_secedn	.577 (.135)	.000	1.781	1.366-2.321
Mother_hiedn	.354 (.116)	.002	1.425	1.136-1.787
Sex of Child				
Child_boy^R				
Child_girl	.075 (.055)	.172	1.078	.968-1.202
Sex of Household				
Male^R				
Female	-.275 (.081)	.001	.759	.648-.890
Father's Education				
Father_illiterate^R				
Father_elemedn	.310 (.123)	.012	1.364	1.071-1.736
Father_secedn	.382 (.126)	.002	1.465	1.145-1.875
Father_hiedn	.108 (.102)	.292	1.114	.911-1.362
Place of Residence				
Rural	.155 (.078)	.047	1.168	1.002-1.360
Use of Contraceptive				
Trad_method^R	.100 (.100)	.317	1.105	.909-1.343
Mode_method	.267 (.109)	.014	1.307	1.055-1.617
Others_method	.075 (.091)	.410	1.078	.902-1.288
Wealth Index				
Poorest^R				
Poor	1.103 (.128)	.000	3.013	2.346-3.868
Middle	.836 (.122)	.000	2.307	1.818-2.928
Richer	.561 (.120)	.000	1.752	1.385-2.217
Variable	β	Sig.	Odd ratio	95% C.I.
Richest	.348 (.122)	.004	1.416	1.116-1.798
Body Mass Index (BMI)				
BMI ≤18.5 kg/m2^R				
BMI ≥18.5 kg/m2	.954 (.080)	.000	2.597	2.218-3.040
Breastfeeding				
Within 1st hour^R				
More than 1st hour	-.187 (.055)	.001	.829	.744-.924

Table 4. (Continued)

Variable	β	Sig.	Odd ratio	95% C.I.
Mother's Age Cohort				
15-19R				
20-24	-.495 (.443)	.264	.610	.256-1.452
25-29	-.222 (.354)	.530	.801	.400-1.603
30-34	-.031 (.352)	.929	.969	.486-1.932
35-39	.033 (.354)	.927	1.033	.517-2.066
40-44	-.069 (.360)	.849	.934	.461-1.893
45-49	-.023 (.387)	.953	.978	.458-2.087
N	41751			
Cox & Snell R Square	0.087			
Nagelkerke R Square	0.116			
-2 Log-likelihood	7595.278			

R, reference category/
Figures in the parentheses are SE of estimates/

There is a positive association between use of contraceptive and wasting. The prevalence of wasting is lower in those families who are using modern method of contraceptive (OD = 1.307, SE = .109).

The forgoing analysis depicts that income is an important variable which determines wasting. The results confirm that wealth index of household has a strong and positive effect on reducing wasting in children. As expected, poor households do not provide nutritional food and better health and medical facilities to their children which result in a high rate of death and mortality. Due to low purchasing power, poor households do not afford the cost of expenditure and improved sanitation.

Here, BMI is used as a proxy of nutritional status of mother which has a significant and positive effect on wasting with high odd ratio (OD = 2.597, SE = 089). The coefficient is statistically significant at 1 percent level of significance.

In this model, mothers age cohort and sex of the child do not have a significant effect on wasting.

The logit estimates of Table 5 demonstrate that the effect of mother's education on child underweight is negative, with the high odds ratio more than one in each level of education and coefficients are statistically significance at 1 percent levels of significance.

Table 5. Determinants of underweight using multiple logistic regression results

Variable	B	Sig.	Odd ratio	95% C.I.
Mother's Education				
Mother_illit[R]				
Mother_elemedn	.685 (.136)	.000	1.984	1.518-2.592
Mother_secedn	.567 (.146)	.000	1.764	1.325-2.348
Mother_hiedn	.404 (.126)	.001	1.497	1.171-1.915
Sex of Child				
Child_boy[R]				
Child_girl	.057 (.058)	.323	1.059	.945-1.187
Sex of Household				
Male[R]				
Female	-.089 (.085)	.296	.915	.774-1.081
Father's Education				
Father_illiterate[R]				
Father_elemedn	.377 (.131)	.004	1.459	1.128-1.886
Father_secedn	.313 (.134)	.019	1.368	1.052-1.780
Father_hiedn	.160 (.111)	.149	1.173	.944-1.457
Place of Residence				
Rural	.143 (.082)	.081	1.154	.983-1.354
Use of Contraceptive				
Trad_method[R]	.133 (.105)	.204	1.142	.930-1.402
Mode_method	.395 (.113)	.000	1.485	1.189-1.854
Others_method	.037 (.096)	.699	1.038	.860-1.252
Wealth Index				
Poorest[R]				
Poor	.941 (.136)	.000	2.562	1.961-3.348
Middle	.728 (.131)	.000	2.072	1.604-2.677
Richer	.542 (.129)	.000	1.719	1.334-2.215
Richest	.460 (.131)	.000	1.584	1.224-2.049
Body Mass Index (BMI)				
BMI $\leq$ 18.5 kg/m2[R]				
BMI $\geq$ 18.5 kg/m2	-1.657 (.083)	.000	.191	.162-.225
Breastfeeding				
Within 1st hour[R]				
More than 1st hour	-.015 (.065)	.821	.985	.867-1.120

Table 5. (Continued)

Variable	B	Sig.	Odd ratio	95% C.I.
Mother's Age Cohort				
15-19[R]				
20-24	-.466 (.469)	.320	.628	.251-1.573
25-29	-.166 (.368)	.652	.847	.411-1.743
30-34	.034 (.366)	.927	1.034	.505-2.118
35-39	.023 (.367)	.949	1.024	.498-2.102
40-44	.042 (.374)	.910	1.043	.501-2.172
45-49	.251 (.402)	.533	1.285	.584-2.826
N	41751			
Cox & Snell R Square	0.118			
Nagelkerke R Square	0.159			
-2 Log-likelihood	6973.408			

R, reference category.
Figures in the parentheses are SE of estimates.

The rates of underweight were lower among children whose mothers have elementary (OD = 1.984, SE = .136), secondary (OD = 1.764, SE = .146) and higher level of education (OD = 1.497, SE = .126) in comparison to mothers with no education. This result is supported by many other studies in the domain of health and education research which prove that educated and qualified mothers are contributing factors for reducing malnutrition (Frosta et al., 2005; Basu & Stephenson, 2005; Miller & Rodgers, 2009; Cleland, 2010; Bbaale, 2014). Waihenya et al. (1996) described that children whose mothers have received elementary education they lower malnutrition rate in comparison to those whose mothers have completed higher education, especially in case of stunting.

The effect of father's education on children underweight at (OD = 1.459, SE = .131), secondary (OD = 1.368, SE = .134) level is found to be statistically significant, with the high odd ratio. However, our result is unfavourable in case of fathers who have earned higher education. It suggests that there is essential to provide education to the fathers up to the secondary level for dipping underweight in children.

Girl children have higher relative odds of 1.059 for underweight in comparison with their boy counterparts and though, this association is not statistically significant. In our society, it is hypothesized that girl children are supposed to be nutritionally ignored due to their low status in society and hence are expected to be more underweight. No significant association is found between sex of household and underweight. However, a significant association is observed between location of household and underweight. Children residing in rural areas (OD = 1.154, SE = .082) are at higher risk of underweight as that of children residing in urban areas, however, the coefficient is statistically significant at 10 percent level of significance.

The estimated results indicate a significant and robust association between family using the modern method of contraceptive and underweight. The rates of underweight in children were lower whose family is using modern method of contraceptive (OD = 1.485, SE = .113) in reference to a family not using any contraceptive.

Similarly, a significant association is observed between the wealth status of the households and underweight. Higher odds of underweight were observed among children who belonged to poor wealth quartile (OD = 2.562, SE = .131) than that of the richest income quartile. Household's level of living also determine the children's nutritional status as result shows that children from the poorest household are more likely to malnourished than children from the richer and richest wealth quintile.

The findings of logit regression demonstrate that children whose mother's BMI is ≥18.5 have statistically significant at 1 percent level of significance with a low odds ratio (OD = .091, SE = .083). According to NFHS-4 report, the percentage of women whose BMI was below normal (BMI < 18.5 $kg/m^{2)}$ was 25.3 percent and children age 6-59 months who were anemic, was 63.2 percent. It is appalling to note that in Uttar Pradesh, more than half of pregnant women (51%) aged 15-49 years were anemic in 2016. However, in out result there is no significant association found between mother's age cohort and underweight in children.

CONCLUSION

This chapter examined the effect of socio-economic factors on child malnutrition status in Uttar Pradesh using data from NFHS-4. This study accounts for maternal level information such as the educational level of mother, malnutrition of mother (BMI as a proxy) and age cohort of the mother that appears to cause malnutrition in children. Among socio-economic determinants, mother's education beyond elementary level emerged as a significant factor that influenced stunting, wasting and underweight of children belonging to the age of five. This result was buttressed by other empirical studies and their results confirmed that mother's educational level beyond secondary level reduces the child nutrition outcomes (Bbaale, 2014). However, the result was not supportive in case of father's educational level more than secondary. It indicated that father's educational level up to secondary was necessary to lower the rate of stunting, wasting and underweight in children. The analysis revealed that sex of the household, sex of the child and mother's age cohort no longer affected the rate of stunting, wasting and underweight of children. However, place of residence determines a significant and positive effect on nutritional status of the children. It means if children reside in rural areas, they were more wasted in reference to those who were residing in urban areas as given the fact that urban settings usually have improved ailment prevention forces. Recently, Government of India has introduced two major health programmes, for instance, Ayushman Bharat Pradhan Mantri Arogya Yojana (PM-JAY) in September 2018, in order to provide primary, secondary and tertiary level treatments to the poor and vulnerable and secondly, National Rural Health Mission in April, 2005 with the aim of enlightening the accessibility of and access to universal health care for individuals living in rural areas. The study finds that wealth index is also more likely to be an imperative factor which exhibits a lower level of child malnutrition. It also represents if households belonging to the richest quintile of wealth index, children's live better quality of life and low mortality.

On the basis of our findings, the present study suggests that malnutrition in children is the result of multiple factors. Low rate of female education is the principal factor for child malnutrition and this happens due to many socio-cultural reasons. To improve the status of health and nutritional outcome, Government of India has launched many programmes, however, more emphasis should also be given to improvement in female education, employment opportunities to enhance economic status, access to rural nutrition, awareness about use of contraceptive. Wealth status is another determinant which indicates high incidence of poverty is a persistent problem affecting the ability to access and attain an adequate diet. This estimate shows the scarcity of material possession and unavailability of basic needs to the poor which results in low demand for energy and protein intake. To improve nutritional status of children, the state needs to eradicate all forms of poverty and to provide employment opportunities to the poor. The study recommends that efforts to improve female education need to be combined with more specific schemes to progress and better child rearing practices. Also, it is important to recommend that civic programmes to ensure education to female on how to feed infants and safe them from infection, and adequate, accessible health services to avoid and treat infections can collectively reduce malnutrition in children. Our results show that family using modern of contraceptive is associated with positive child health outcomes. Therefore, the present study suggests that government, donors and private players/NGO can perform an imperative role in improving women's access to modern healthcare services by ensuring affordability at the regional level.

REFERENCES

Ahmed, T., Roy. S. & Alam. N., et al. (2012). Determinants of undernutrition in children under 2 years of age from rural Bangladesh. *Indian Pediatric*, 49:821–24.

Allen, L. H., Gillespie, S. R. (2001). *What works?: A review of the efficacy and effectiveness of nutrition intervention. ACC/SCN Nutrition Policy Paper*, 19, 27–8.

Bangladesh Demographic and Health Survey (2007). Dhaka, Bangladesh and Calverton, Maryland [USA]: National Institute of Population Research and Training, Mitra and Associates, and Marco International.

Basu, A. M., & Stephenson, R. (2005). Low levels of maternal education and the proximate determinants of childhood mortality: a little learning is not a dangerous thing. *Social Science & Medicine*, 60(9), 2011-2023.

Bbaale, E. (2014). Maternal education and child nutritional status: evidence from Uganda, *African Journal of Economic and Management Studies*, 5(1), 52-74.

Burroway, Rebekah (2016). Empowering women, strengthening children: a multi-level analysis of gender inequality and child malnutrition in developing countries. *Gender and Food: From Production to Consumption and After*, 22, 117-142.

Caldwell, J. (1979). Education as a factor in mortality decline: an examination of Nigerian Data. *Population Studies*, 33(3), 395-413.

Case, A. (2001). Does money protect health status? Evidence from South African pensions, *NBER working paper no. 8495*.

Case, A., Lubotsky, D. & Paxson, C. (2002). Economic status and health in child: the origins of the gradients. *American Economic Review*, 95(5), 1308-1334.

Choudhury, P. K. (2015). Explaining the role of parental education in the regional variations in infant mortality in India. *Asia & the Pacific Policy Studies*, 2(3), 544-572.

Cleland, J. (2010). The benefits of educating women. *The Lancet*, 376(9745), 933-934.

Das, Sibabrata & Sahoo, Harihar (2011). An investigation into factors affecting child undernutrition in Madhya Pradesh. *Anthropologist*, 13(3): 227-233.

Dasgupta, M., Lokshin, M., Gragnolati, M. & Ivaschenko, O. (2005) *Improving Child Nutrition Outcome in India-Can the Integrated Child*

Development Services Program be More Effective? World Bank Policy Research Working Paper No. 3647. World Bank, Washington, DC.

Desai, S., & Alva, S. (1998). Maternal education and child health: is there a strong casual relationship? *Demography*, 35(1), 71-81.

Deshmukh, P. R., Sinha, N. & Dongre, A. R. (2013). Social determinants of stunting in rural area of Wardha, Central India. *Medical Journal Armed Forces India*, 69, 213-217.

Frost, M. B., Forste, R., & Haas, D. W. (2005). Maternal education and child nutritional status in Bolivia: finding the links. *Social Science & Medicine*, 60(2), 395-407.

Glewwe, P. (1999). Why does mother's schooling raise child health in developing countries? Evidence from Morocco. *Journal of Human Resources*, 34(1), 124-159.

Govindasamy, P. & Ramesh, B. M. (1997). Maternal education and utilisation of maternal and child health services in India. *NFHS subject reports, Number 5, International Institute of Population Sciences*, Mumbai.

Grantham-McGregor, S., Cheung, Y. B., Glewwe, P., Richter, L., Strupp, B. (2007). Developmental potential in the first 5 years for children in developing countries. *Lancet*, 369 (9555): 60-70.

Haddad, L., Alderman, H., Appleton, S., Song, L., Yohannes, Y. (2003). Reducing child malnutrition: how far does income growth take us? *The World Bank Economic Review* 17 (1), 107–131.

Headey, D., (2013). Developmental drivers of nutritional change: a cross-country analysis. *World Development*, 42, 76–88.

Heltberg, R., (2009). Malnutrition, poverty, and economic growth. *Health Econ*. 18 (S1), S77–S88.

Horton, S. (1998). Birth order and child nutritional status: evidence from the Philippines. *Economic Development and Cultural Change*, 36(2), 341-354.

Mazumdar, S. (2010). Determinants of inequality in child malnutrition in India, *Asian Population Studies*, 6(3), 307-333.

Measham, A. R. & Chatterjee, M. (1999). *Wasting away: The crisis of malnutrition in India.* Washington, DC: World Bank.

Miller, J. E., & Rodgers, Y. (2009). Mother's education and Children's nutritional status: new evidence from Cambodia. *Asian Development Review*, 26(1), 131-165.

Miller, J. E., & Rodgers, Y. V. (2009). Mother's Education and Children's Nutritional Status: New Evidence from Cambodia, *Asian Development Review*, 26(1), 131-165.

Mishra, V. K., Lahiri, S., & Luther, N. Y. (1999). *Child nutrition in India. National Family Health Survey Subject Reports 14* (June 1999). Mumbai, India: International Institute for Population Sciences, and Honolulu, Hawaii, USA: East-West Center.

Mondal, N. I., Hossain, K., Ali, K. (2009). Factors influencing infant and child mortality: a case study of Rajshahi district, Bangladesh. *Journal of Human Ecology*, 26(1), 31-39.

Moore, A C., Akhter, S., & Aboud, F. E. (2006). Responsive complementary feeding in rural Bangladesh. *Social Science & Medicine*, 62, 1917–30.

Nussbaum, M. (2004). Women's education: A global challenge. *Signs: Journal of Women in Culture and Society*, 29(2), 325-355.

Patel, A., Badhoniya, N., Khadse, S., Senarath. U., Agho, K. E., & Dibley, M. J. (2010). Infant and young child feeding indicators and determinants of poor feeding practices in India: secondary data analysis of National Family Health Survey 2005-06. *Food and Nutrition Bulletin*, 31, 314-33.

Sastry, N., (1997). What explains rural-urban differentials in child mortality in Brazil? *Social Science and Medicine*, 44(7), 989-1002.

Sen, A. (1999). *Development as freedom*. New York, NY: Anchor Books.

Sethuraman, K., Lansdown, R. & Sullivan, K. (2006). Women's empowerment and domestic violence: The role of sociocultural determinants in maternal and child undernutrition in tribal and rural communities in South India, *Food and Nutrition Bulletin,* 27(2), 128-143.

Shetty, P. S. & James, W. P. T. (1994). Body mass index: a measure of chronic energy deficiency in adults. Rome, Italy: Food and Agriculture

Organization of the United Nations, *(FAO food and nutrition paper 56)*.

Smith, L. C., & Haddad, L. J., (2000). *Explaining Child Malnutrition in Developing Countries, Technical report*. International Food Policy Research Institute (IFPRI), Washington, DC.

Subramanyam, M. A., Kawachi, I., Berkman, L. F. & Subramanian, S. V. (2010). Socioeconomic inequalities in childhood undernutrition in India: Analyzing trends between 1992 and 2005. *PLoS One,* 5(6), e11392.

Svedberg, P. (2008). Why malnutrition in shining India persist. *4th Annual Conference on Economic Growth and Development*, New Delhi, 2008.

Tarozzi, Alessandro & Mahajan, Aprajit (2007). Child nutrition in India in the nineties. *Economic Development and Cultural Change*, University of Chicago Press, 55(3), 441-486.

Thomas, D., Strauss, J., Henriques, M. H. (1991). How does mother's education affect child health? *Journal of Human Resources*, 26(2), 183-211.

United Nations (2019). *Transforming Our World: The 2030 Agenda for Sustainable Development*. https://sustainabledevelopment.un.org/post 2015/transformingourworld.

Waihenya, E. W., Kogi-Makau, W. & Muita, J. W. (1996). Maternal nutritional knowledge and the nutritional status of preschool children in a Nairobi slum. *East African Medical Journal*, 73, 419-423.

Walker, S. P., Wachs, T. D., Gardner, M., Lozoff, B., Wasserman, G. A., Pollitt, E. & Carter, J. A. (2007). Child development: risk factors for adverse outcomes in developing countries. *The Lancet,* 369(9556), 145-157.

Wemakor, A., Garti, H., Azongo, T., Garti, H., & Atosona, A., (2018). *Young maternal age is a risk factor for child undernutrition in Tamale Metropolis, Ghana*. BMC Research Note, 11: 877.

WHO (1978). *A Growth Chart for International Use in Maternal and Child Health Care*. WHO: Geneva.

WHO (2010). *Nutrition Landscape Information System (NLIS) Country Profile Indicators*. Geneva, Switzerland.

World Bank (2006). *Repositioning Nutrition as Central to Development A Strategy for Large-Scale Action*. Washington DC, USA.

APPENDIX

Table A-1. District and sex wise nutritional status of under- five children in Uttar Pradesh

Sex of Child	District	n	Underweight	Rank	N	Stunted	Rank	n	Wasted	Rank
Male	Saharanpur	120	33.7	57	126	35.5	67	63	17.7	41
	Muzaffarnagar	168	38.4	40	184	42.1	55	98	22.4	23
	Bijnor	161	42.9	26	164	43.7	50	89	23.7	16
	Moradabad	216	44.2	23	229	46.9	29	92	18.8	38
	Rampur	109	43.4	25	109	43.4	51	53	21.1	29
	Jyotiba Phule Nagar	86	45.5	19	88	46.6	30	43	22.8	21
	Meerut	168	34.1	54	174	35.4	68	93	18.9	37
	Baghpat	46	33.8	55	54	39.7	61	25	18.4	40
	Ghaziabad	153	30.7	66	191	38.4	65	70	14.1	57
	Gautam Buddha Nagar	66	29.1	68	74	32.5	71	39	17.1	46
	Bulandshahr	133	35.5	48	165	43.9	48	64	17.0	47
	Aligarh	150	37.4	43	197	49.1	22	65	16.2	48
	Mahamaya Nagar	41	28.3	70	62	42.8	53	13	9.0	71
	Mathura	78	27.7	71	121	42.9	52	38	13.5	60
	Agra	146	34.5	51	195	46.1	33	63	14.9	53
	Firozabad	73	29.8	67	114	46.5	31	32	13.1	62
	Mainpuri	60	36.6	44	78	47.9	28	23	14.1	56
	Budaun	239	54.6	1	253	57.6	7	92	21.0	30
	Bareilly	153	38.4	39	179	44.9	44	82	20.6	31
	Pilibhit	98	50.3	5	108	55.4	8	40	20.5	32
	Shahjahanpur	157	48.8	7	148	46.0	35	78	24.3	14
	Kheri	165	35.7	47	241	52.3	14	86	18.6	39
	Sitapur	228	45.9	17	288	58.1	6	61	12.3	63

Sex of Child	District	n	Underweight	Rank	N	Stunted	Rank	n	Wasted	Rank
	Hardoi	173	41.9	31	216	52.3	13	62	15.0	52
	Unnao	105	35.4	49	134	45.1	42	45	15.1	51
	Lucknow	103	44.4	22	90	38.8	64	82	35.3	2
	Rae Bareli	109	42.1	30	103	39.6	62	79	30.5	8
	Farrukhabad	56	30.9	64	87	48.1	26	18	9.9	69
	Kannauj	46	30.9	65	74	50.0	17	18	12.1	64
	Etawah	53	33.8	56	82	52.6	11	21	13.4	61
	Auraiya	53	44.9	21	52	44.1	47	35	29.4	9
	Kanpur Dehat	60	41.7	33	66	45.8	36	33	22.8	20
	Kanpur Nagar	136	42.2	29	141	43.8	49	78	24.2	15
	Jalaun	57	47.5	10	49	40.8	58	40	33.3	5
	Jhansi	57	39.9	37	48	33.6	70	45	31.7	7
	Lalitpur	44	46.8	12	43	45.7	37	33	35.1	3
	Hamirpur	30	38.5	38	32	41.0	56	28	35.4	1
	Mahoba	34	48.6	8	31	44.3	46	20	28.6	11
	Banda	46	41.8	32	50	45.5	40	24	21.8	27
	Chitrakoot	55	52.4	3	55	52.4	12	34	32.4	6
	Fatehpur	69	37.7	42	89	48.6	23	32	17.5	44
	Pratapgarh	110	46.2	15	101	42.4	54	63	26.4	13
	Kaushambi	77	54.2	2	70	49.3	20	48	33.8	4
	Allahabad	245	45.9	16	238	44.7	45	123	23.0	18
	Bara Banki	120	45.1	20	141	53.0	10	37	14.0	59
	Faizabad	99	47.1	11	113	53.6	9	48	22.9	19
	Ambedkar Nagar	91	42.7	28	98	46.0	34	47	22.1	26
	Sultanpur	131	42.8	27	138	45.0	43	59	19.2	34
	Bahraich	189	47.8	9	262	66.5	1	58	14.7	55
	Shrawasti	51	37.8	41	82	60.7	2	16	11.8	65
	Balrampur	90	41.3	34	132	60.6	3	23	10.6	66
	Gonda	127	40.2	36	184	58.2	4	33	10.4	68
	Siddharth Nagar	121	44.2	24	160	58.2	5	43	15.6	49
	Basti	87	31.2	62	126	45.2	41	49	17.6	43
	Sant Kabir Nagar	62	34.3	52	89	49.2	21	19	10.5	67
	Mahrajganj	121	40.6	35	154	51.7	16	44	14.8	54
	Gorakhpur	107	28.9	69	148	39.9	60	71	19.1	35
	Kushinagar	149	36.5	45	186	45.6	39	57	14.0	58
	Deoria	85	31.0	63	106	38.8	63	52	19.0	36
	Azamgarh	110	31.7	61	126	36.4	66	61	17.6	42

Table A-1. (Continued)

Sex of Child	District	n	Underweight	Rank	N	Stunted	Rank	n	Wasted	Rank
	Mau	68	34.2	53	81	40.9	57	42	21.2	28
	Ballia	100	32.6	58	108	35.1	69	53	17.3	45
	Jaunpur	206	51.9	4	198	49.9	18	117	29.4	10
	Ghazipur	95	34.8	50	125	45.6	38	61	22.3	24
	Chandauli	81	35.8	46	90	40.0	59	46	20.4	33
	Varanasi	136	45.8	18	138	46.5	32	83	27.9	12
	Sant Ravidas Nagar (Bhadohi)	83	49.7	6	87	51.8	15	37	22.2	25
	Mirzapur	101	46.3	14	105	48.4	24	51	23.5	17
	Sonbhadra	88	46.6	13	90	47.9	27	43	22.8	22
	Etah	59	31.9	59	91	49.5	19	18	9.7	70
	Kanshiram Nagar	51	31.9	60	77	48.1	25	25	15.6	50
	Total	7540	39.7		8828	46.5		3656	19.2	
Female	Saharanpur	129	40.2	32	125	38.9	61	63	19.6	24
	Muzaffarnagar	126	36.6	45	139	40.3	58	57	16.5	33
	Bijnor	121	40.7	30	122	41.1	57	66	22.2	17
	Moradabad	188	42.0	25	197	44.0	44	64	14.3	44
	Rampur	92	46.2	14	93	46.7	31	43	21.6	21
	Jyotiba Phule Nagar	71	40.1	34	75	42.4	52	39	22.2	18
	Meerut	157	37.0	44	149	35.2	66	83	19.6	25
	Baghpat	35	34.0	59	33	32.0	70	11	10.6	58
	Ghaziabad	130	28.8	69	142	31.4	71	63	13.9	46
	Gautam Buddha Nagar	57	29.7	67	63	32.8	69	30	15.6	37
	Bulandshahr	115	33.3	62	153	44.3	40	53	15.4	40
	Aligarh	141	39.5	36	175	49.2	25	48	13.4	47
	Mahamaya Nagar	50	36.0	47	64	46.0	36	14	10.1	60
	Mathura	68	27.6	70	92	37.4	63	27	11.0	55
	Agra	145	35.0	53	181	43.7	47	58	14.0	45
	Firozabad	54	25.7	71	87	41.4	56	22	10.5	59
	Mainpuri	42	30.4	66	62	45.3	38	9	6.6	71
	Budaun	220	53.5	3	214	52.1	18	81	19.8	23
	Bareilly	191	44.7	19	199	46.6	33	74	17.3	29
	Pilibhit	72	38.7	40	91	49.2	24	45	24.3	9
	Shahjahanpur	195	60.2	1	174	53.5	13	76	23.5	12

Sex of Child	District	n	Underweight	Rank	N	Stunted	Rank	n	Wasted	Rank
	Kheri	189	47.4	11	224	56.1	6	67	16.8	31
	Sitapur	177	51.0	8	183	52.7	15	54	15.6	39
	Hardoi	126	37.7	42	162	48.4	28	48	14.4	43
	Unnao	96	35.4	50	132	48.7	27	30	11.1	53
	Lucknow	86	42.6	23	68	33.7	68	63	31.2	4
	Rae Bareli	100	41.0	29	85	34.8	67	80	32.7	3
	Farrukhabad	61	33.5	61	91	50.0	21	13	7.1	70
	Kannauj	56	34.6	57	84	51.5	19	20	12.3	48
	Etawah	42	33.3	63	69	54.8	10	14	11.1	54
	Auraiya	46	45.1	18	45	44.1	43	24	23.5	11
	Kanpur Dehat	50	34.2	58	68	46.6	34	14	9.6	64
	Kanpur Nagar	127	41.6	54	130	42.6	51	73	23.9	10
	Jalaun	43	53.1	4	44	54.3	11	24	29.6	6
	Jhansi	47	39.5	37	47	39.5	60	26	21.8	20
	Lalitpur	48	51.6	7	33	35.5	65	40	42.6	1
	Hamirpur	35	45.5	17	29	37.2	64	24	31.2	5
	Mahoba	30	46.2	15	29	43.9	45	12	18.5	27
	Banda	45	42.9	21	49	46.7	32	16	15.2	41
	Chitrakoot	52	52.0	6	49	49.0	26	34	34.0	2
	Fatehpur	66	42.6	22	87	56.1	7	18	11.6	50
	Pratapgarh	88	40.2	33	91	41.6	55	50	22.8	14
	Kaushambi	72	50.3	9	72	50.3	20	39	27.3	7
	Allahabad	213	42.3	24	217	43.1	49	82	16.3	34
	Bara Banki	103	36.5	45	140	49.6	23	30	10.6	57
	Faizabad	88	44.4	20	94	47.5	29	31	15.7	36
	Ambedkar Nagar	64	37.9	40	67	39.6	59	38	22.5	15
	Sultanpur	88	34.8	55	112	44.3	42	46	18.1	28
	Bahraich	155	39.4	38	255	64.9	1	47	11.9	49
	Shrawasti	47	40.5	31	73	62.9	3	11	9.5	66
	Balrampur	94	45.9	16	133	64.9	2	20	9.8	63
	Gonda	123	36.0	47	189	55.3	9	27	7.9	68
	Siddharth Nagar	104	41.6	27	142	56.6	4	29	11.6	51
	Basti	84	35.7	48	125	53.4	14	23	9.8	62
	Sant Kabir Nagar	64	40.0	35	84	52.5	16	18	11.3	52
	Mahrajganj	86	32.8	63	148	56.5	5	25	9.5	65
	Gorakhpur	144	41.3	28	153	43.8	46	69	19.8	22
	Kushinagar	124	35.2	50	161	45.7	37	55	15.6	38
	Deoria	84	31.6	64	116	43.6	48	24	9.0	67
	Azamgarh	124	35.2	51	148	42.0	54	61	17.3	30
	Mau	67	37.4	42	76	42.2	53	34	19.0	26

Table A-1. (Continued)

Sex of Child	District	n	Underweight	Rank	n	Stunted	Rank	n	Wasted	Rank
	Ballia	82	29.2	68	125	44.3	41	30	10.7	56
	Jaunpur	180	54.4	2	153	46.1	35	83	25.1	8
	Ghazipur	86	29.4	67	110	37.5	62	43	14.7	42
	Chandauli	66	34.9	54	89	47.1	30	30	15.9	35
	Varanasi	130	47.3	12	123	44.6	39	62	22.5	16
	Sant Ravidas Nagar (Bhadohi)	83	52.2	5	84	52.5	17	35	21.9	19
	Mirzapur	99	47.6	10	104	50.0	22	35	16.8	32
	Sonbhadra	86	46.7	13	79	42.9	50	43	23.4	13
	Etah	52	33.8	59	83	53.9	12	15	9.8	61
	Kanshiram Nagar	49	35.0	53	78	55.3	8	11	7.9	69
	Total	6850	39.8		7962	46.2		2866	16.7	

INDEX

D

E

F

G

H

I

J

K

L

M

N

O

P

R

S

T

U

V

W

Y

Z